REED'S
INSTRUMENTATION AND
CONTROL SYSTEMS

By
LESLIE JACKSON
B.Sc., C.Eng., F.I.Mar.E., F.R.I.N.A.
EXTRA FIRST CLASS ENGINEERS' CERTIFICATE

THOMAS REED
PUBLICATIONS LTD
Nautical Publishers since 1782
LONDON □ HAMBURG □ BOSTON

First Edition – 1970
Second Edition – 1975
Third Edition – 1979
Reprinted – 1985
Fourth Edition – 1992

ISBN 0 947637 86 9

Printed and bound in Great Britain

PREFACE

This book aims to bridge the gap between the mathematical treatment often used by the specialist control engineer and the (necessarily) narrow descriptive literature of a particular manufacturer.

It is written primarily for those with a good general engineering background who have had little experience in instrumentation and control.

The work favours marine engineering but students and engineers in other industries should find it a useful book as the subject has a common basis. Text and examples cover the requirements of all Department of Transport (DTp) – and all Business and Technician Education Council (BTEC) and SCOTVEC – syllabuses and examinations for marine engineer officers and cadets.

Full use has been made of simplified diagrams and the work is presented from basic principles, using analogues where appropriate.

The introduction is followed by five chapters on variable measurement in instrumentation. Chapters 6–8 on telemetering, electronic devices and final controlling elements link instrumentation to control. Chapters 9–14 cover theory, practice and components of process and kinetic control systems. Chapter 15 and 16 are intended to develop a broader knowledge of the subject and, by necessity, have a more analytic and mathematical approach.

A selection of test examples are included at the end of each chapter and specimen examination questions are added at the end of the book.

If the reader wishes to obtain a full detailed description of a particular component reference should be made to manufacturers' instruction manuals.

CONTENTS

CHAPTER 7 – ELECTRONIC DEVICES

Analogue and digital circuits.

Semi-conductors; atomic theory, electron conduction, solid state junctions, rectifier and transistor.

Rectifiers; bridge and centre tap transformer, applications, carrier, modulation, de-modulation, smoothing, filter. Semi-conductor diode, zener diode, characteristics, voltage stabilisation, applications. Thyristor (silicon controlled rectifier). Triac, diac.

Amplifiers; rotating electrical (metadyne), magnetic (transducer), classification, junction transistor, circuit configurations, transfer characteristics, voltage amplifier, parameters, T-circuit, unijunction transistor (oscillator). Linear mode small signal transistor, power amplifier (push-pull), dc amplifier, chopper, field effect transistor (JFET, MOSFET), operational amplifier. Feedback, negative, positive, gain, stability.

Oscillators; basic theory, transistor oscillator, harmonic oscillators, relaxation oscillators, push-pull blocking, multivibrator, univibrator.

Other devices; digital, switching, logic, analysers, cathode ray oscilloscope. Radio communication, electromagnetic radiation spectrum. Light emitting diode (LED). Fibre optics.   89-114

CHAPTER 8 – FINAL CONTROLLING ELEMENTS

Correcting units; diaphragm control valve, motor, correcting element (valve) characteristics, valve positioner. Piston operated, torque actuated, rotary cylinder and wax element control valves.

Servo-motors; electric dc and ac motors, hydraulic ram servo, variable delivery pump, hydraulic rotary vane servo, pneumatic piston servo, other servos.   115-126

INTRODUCTION

HISTORICAL

Instrumentation has always been an integral part of technology. Development from simple level indicators, Bourdon tubes, moving iron and moving coil meters, etc. has been rapid. Progress in electrics and electronics has led to centralised recording and display stations with associated data processing, computing and control systems. Application to control with the requirement of accurate measuring (sensors), variable converting (transducers) and remote signal transmission (telemetering) has involved a close relation between measuring, processing and control systems. The advantage of electrical signal transmission is apparent in the development of instrumentation. Modular designs and interchangeable plug-in sub-assemblies have improved servicing of electronic units. Digital operation is increasingly being favoured over analogue operation.

The development of control elements is inherent in the history of man as life form itself utilises classic control principles. The Watt governor (1788) was one of the first practical applications. Instability was recognised last century in hunting of steam engine speed, and ship steering gears, and much analysis followed. The ship steering gear remains as one of the first control systems and its development alongside modern bridge equipment continues. By 1939 fairly complex systems, mainly pneumatic and hydraulic, were in use and development in electronics, related to the feedback amplifier, has accelerated progress and now leads to computer control. Systems are generally classified by their field of operation. Process control such as flow, level and pressure; kinetic control such as displacement, velocity and acceleration; etc.

UTILISATION

Automation is essentially sequence-controlled mechanisation and is only an aspect of control. The word tends to varied usage and as such is not considered further in this text. American terminology tends to use the word *cybernetics* to describe the entire field of communication and control theory. A full study programme of the subject, which is generally followed in this text, is shown in the first sketch.

The degree of utilisation in marine practice varies a great deal. Individual control loops, from simple to fairly complex, have been in use for many years. Centralised data handling has been a more recent innovation. Bridge control and unmanned machinery spaces have developed quite rapidly with improved reliability of control and alarm loops. The next stage of development is to link the centralised data handling system to an integrated central control system. This requires that a computer is involved in the interface between measurement and control. Computer control has developed most recently, from small programmed functions to quite sophisticated direct digitally controlled processes. A modern computer can be so programmed not only to control machinery under all conditions but also to have start up, emergency and shut down procedures. Extension to navigational, cargo handling, etc. easily follows.

ECONOMY

Instrumentation and control results in more efficient operation and reduced manpower in every case. There is an increased first cost due to specialised equipment provision which leads to increased insurance requirements and some increase in certain running costs, e.g. staff training, skilled maintenance, etc. Overall running costs are reduced because of large cost savings in fuel and general maintenance, due to efficient operation and close supervision, as well as staff reductions. The annual saving taking all factors into account is well proven for controlled plant and the factor increases with increasing size of plant and machinery.

SAFETY

In most cases safety is improved by monitoring and control. Reliability of both measuring-control alarm equipment is essential

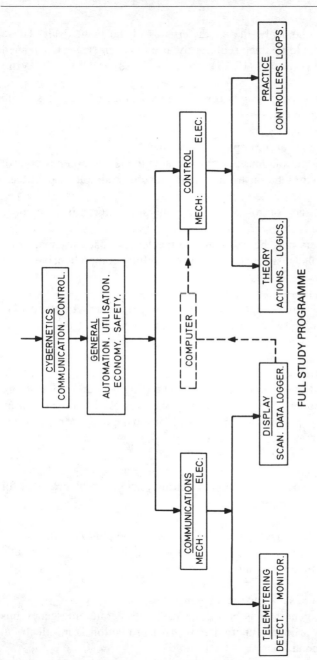

FULL STUDY PROGRAMME

as indeed should be the characteristics of the plant itself. These aspects should be analysed closely at the design stage. Unmanned machinery spaces (UMS) are now classified by surveying societies.

Essential requirements for unattended machinery spaces could be summarised thus:

1. Bridge control of propulsion machinery.

The bridge watchkeeper must be able to take emergency control action. Control and instrumentation must be as simple as possible.

2. Centralised control and instruments are required in machinery space.

Engineers may be called to the machinery space in emergency and controls must be easily reached and fully comprehensive.

3. Automatic fire detection system.

Alarm and detection system must operate very rapidly. Numerous well sited and quick response detectors (sensors) must be fitted.

4. Fire extinguishing system.

In addition to conventional hand extinguishers a control fire station remote from the machinery space is essential. The station must give control of emergency pumps, generators, valves, ventilators, extinguishing media, etc.

5. Alarm system.

A comprehensive machinery alarm system must be provided for control and accommodation areas.

6. Automatic bilge high level fluid alarms and pumping units.

Sensing devices in bilges with alarms and hand or automatic pump cut in devices must be provided.

7. Automatic start emergency generator.

Such a generator is best connected to separate emergency bus bars. The primary function is to give protection from electrical blackout conditions.

8. Local hand control of essential machinery.

9. Adequate settling tank storage capacity.

10. Regular testing and maintenance of intrumentation.

TERMINOLOGY

The *detecting element* responds directly to the value of the variable. A *measuring element* responds to the signal from the detecting element and gives a signal representing the variable value. For example pressure (variable), Bourdon tube (detecting element) and linkage pointer, scale, *i.e.* pressure gauge (*measuring element*). *The measuring unit* comprises detecting element and measuring element. Such a unit is used as a *monitoring element* (to convert, when necessary, the actual variable value into a converted variable value) of a process control system

Sensor (American) is a term used for the detecting element. Is, by its very nature, essentially a *transducer*.

Transducer (American) is a device to convert a signal (representing a physical quantity) of one form into a corresponding signal of another form, retaining the amplitude variations of energy being converted. For example a microphone is a sound transducer (acoustic to electrical) and a loud speaker an electrical transducer (electrical to acoustic). A transducer may be an integral part of the measuring unit, for example pressure to displacement in a Bourdon pressure gauge. It may also be a separate unit converter especially suitable to change the signal to a better form for remote transmissions, e.g. displacement to electrical in a differential transformer.

Telemetering may be defined as signal transmission over a considerable distance. In measurement this involves information transfer from detecting element to a central recording-display station. In control this involves control operating devices and related signal transfers. In telemetering systems the measuring unit is often called the *transmitter*, usually incorporating a transducer, and the recording unit some distance away is then referred to as the *receiver* which may have an associated transducer if required.

The terminology involved further to the above and especially related to control systems is now fairly extensive. Such terminology is covered in some detail at the start of Chapter 9.

Instrumentation in this book is generally confined to dynamic systems related to recording and control. Obviously the range is much wider if extended to include static-laboratory type instrument devices.

COMPARISON OF SYSTEMS

Systems, telemetering or control, may be either pneumatic, hydraulic, or electronic-electric, or a combination.

Hydraulic systems are generally more restricted in application. Basically the technique is as for pneumatics but fluid cannot be allowed to escape and a recovery-storage system is required. General use is in the higher pressure range.

A combination of electronic measure-record instrumentation and pneumatic final power control element is very effective. Controllers may be either pneumatic or electronic. The former have generally been used because of proven reliability and ease of application to final power transmission. Electronic controllers are increasingly being used and the all electronic-electric system has many obvious advantages.

The advantage of a pneumatic system are:
1. Less expensive initially, this is in spite of tubing and air supply costs.
2. No heat generation and safe in explosive atmosphere.
3. Less susceptible to power supply variation, but do have appreciable time lags.
4. Direct application, without transducers, to large final power actuators.

The advantages of an electronic system are:
1. Small and adaptable with cheap flexible transmission lines.
2. No moving parts, can however generate heat.
3. Stable, generally accurate and very short time lags.
4. Low power consumption, direct application to computers, but often need final element transducers.

In every case equipment must be robust, reliable, interchangeable, simple and resistant to environment. A long commissioning time should be applied and regular skilled maintenance is required. Signal dc transmission is usually preferred although ac signals are essential for certain variables and easy amplification of ac is an inherent advantage.

CONTROL LOOPS

An open loop system has no feedback and controller action is not related to final result. Consider a domestic central heating system as an example, in which fuel supply is varied manually or automatically by external ambient temperature. Room temperature will be maintained at a reasonable value related to outside conditions. However room temperature does not control fuel supply so that this is open loop. The word loop is really a misnomer.

Now to the open loop shown, add human operator, so closing the loop (dotted lines on sketch). This is a manually controlled closed loop system.

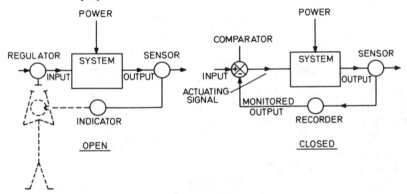

CONTROL LOOPS

The automatically controlled closed loop system is actuated by a signal dependent upon deviation (error) between input (set) and output values. Deviation only exists when monitored output (negative feedback) differs from input and this signal controls power supplied to output. For a closed loop system, as sketched, output power is only controlled by, and not supplied by, the actuating signal. Closed loops have a self regulating property.

SYSTEM

An assembly of linked components within a boundary. The motor car is a good example; mechanical, electrical, control and suspension sub-systems within a body-chassis boundary. A system may have one input and a related output dependent on the effect of that system (**transfer function *G***).

$$\theta_0 = G\theta_i$$

The boundary, represented as a"black box", may include a complex system which need not be analysed if G is provided. More complex systems have interconnecting links to related systems. A system must have input, process, output, and in most systems a source of power and a means of control.

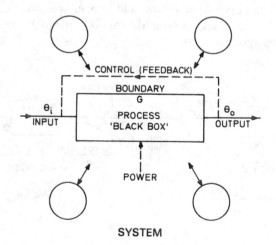

SYSTEM

ANALOGUE

Many different physical phenomena behave in a similar way, i.e. are analogues of each other. Two examples are air escape from a storage vessel and electrical charge loss from a capacitor.

Rate of change of pressure α pressure
Rate of change of charge α voltage

i.e. Rate of change of variable $x = -kx$

$$\frac{dx}{dt} = -kx$$

A solution, where C is the x value when time is zero, is:

$$kt = l_n\frac{C}{x}$$

Voltage (or current) can readily be made analogous to physical phenomena. The traditional electric clock is analogue *i.e.* continuous representation.

DIGITAL

A digital device manipulates "bits", *i.e.* discrete items of information - illustrated by the digital clock representation. States are on/off, equal/unequal, etc. and the binary digit system is utilised.

COMPUTERS

The electronic analogue computer is essentially a simulator on which electrical analogues of various systems can be analysed and illustrated. The digital computer is a machine for routine, repetitive arithmetic. Hybrid types are a combination.

MICROPROCESSORS AND MICROCOMPUTERS

First generation computers (approx. room size) were often, for the same capacity, replaced by minicomputers (say desk size) and in turn by microcomputers (hand size and smaller) following the silicon chip and integrated circuit (IC) development.

A microprocessor (μP) – component product – may be a single chip unit or a collection of unassembled processor-related components such as central processor unit (CPU), timers, memories, interfaces, etc.

A microcomputer (μCP) – board product – assembly of μP components mounted on a printed circuit board sufficient to make up a working computer.

Semi-conductor and metal oxide silicon (MOS) – random access memory (RAM) and read only memory (ROM) are utilised for both CPU and logic system μP.

Whilst the μP, consisting of arithmetic logic units plus sets of registers and control circuits, cannot be used by itself to create a system it can, with support chips, form a dedicated controller or μCP.

CHAPTER 1

TEMPERATURE MEASUREMENT

This chapter is concerned with the practical aspects of thermometry. No consideration is applied to absolute standards or to the consideration of special techniques related to extreme temperatures , etc.

MECHANICAL THERMOMETRY

LIQUID IN GLASS THERMOMETER

Mercury can be used from −38°C (its freezing point) to about 600°C. For the higher temperatures an inert gas at high pressure is introduced as the boiling point of mercury is about 360°C at atmospheric pressure; special glass is also required.

Alcohol is used in the range −80°C to 70°C (or toluene) and pentane can be used to −196°C.

Total immersion types are most accurate, especially when the fluid is coloured and magnification is used. In many cases only temperature *differential* is required so that relatively low accuracy partial immersion types are often satisfactory.

FILLED-SYSTEM THERMOMETER

Consist of liquid, vapour or gas filled types. All utilise a bulb, connecting capillary and usually a Bourdon tube mechanism, responding to pressure change from volume variation (liquid), for pointer or pen operation. Some systems incorporate a compensating capillary and bourdon tube to allow for changes of ambient temperature. Alternatively a bi-metallic link for compensation can be incorporated into the mechanism.

A common type of liquid filled system utilises mercury in steel which can be pressurised for high temperature duty to 600°C. Such devices are most useful for remote sensing and telemetering back to a central instrumentation panel. Capillary bore is about 0·02 mm, the scale is generally linear but calibration must allow for heat variation. Power is sufficient for pointer, pen or transducer operation.

Vapour pressure thermometers commonly use freon, alcohol or ether which partly fills the system as liquid, and the remainder is vapour filled. Measurement of vapour pressure gives an indication of liquid surface temperature and is usually used in the range −50°C to 260°C, with the upper limit fixed by the critical temperature of the liquid which must have a low boiling point (high vapour pressure). The scale is non-linear, ambient variations can be neglected but there can be appreciable time lags and the device is not well suited to remote indication.

Gas filled thermometers usually employ nitrogen or helium under high pressure, and pressure is proportional to absolute temperature at constant volume. The usual temperature range is −50°C to 430°C and the scale is linear. Compensation for ambient temperature variation is difficult. When used as a sensor linked to a pneumatic transducer it is a very effective device.

BI-METALLIC THERMOMETER

The principle of operation of bi-metallic devices is that of differential expansion of two different materials rigidly joined together, one on the other, as a strip of bi-metallic material. Figure 1.1 illustrates a typical design usually employed between −40°C and 320°C. Invar (36% Ni, 64% Fe) has a low coefficient of expansion and when welded to a Ni-Mo alloy gives a good bi-metallic strip. The helix coils or uncoils with temperature variation and as one end is fixed the movement rotates shaft and pointer. The range of the instrument is fixed by the materials used.

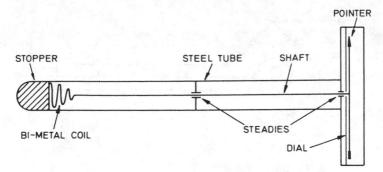

Fig. 1.1 MECHANICAL TYPE THERMOMETER (BI-METALLIC TYPE)

ELECTRICAL THERMOMETRY

RESISTANCE THERMOMETER

The electrical resistance of a metal varies with temperature and this relationship is usually expressed as $\rho_\theta = \rho_0 (1 + \alpha\theta)$
where ρ_θ is the specific resistance at temperature $\theta°$C
    " ρ_0 is the specific resistance at temperature 0°C
    " α is a constant which depends upon the metal and is called the temperature coefficient of resistance.

Figure 1.2 shows diagrammatically a resistance type of temperature measuring unit using the well known Wheatstone bridge principle. r_1 r_2 is a variable resistance used for balance purposes; at balance we have:

$$\frac{R_1 + r_1}{R_2 + r_2} = \frac{R_3 + r}{R_4 + r}$$

r is the resistance of each of the wires and since each wire will be subjected to the same temperature variation along its length their resistances will always be equal.

When the temperature detecting element is subjected to a temperature alteration its resistance alters and the bridge balance is upset. By using the variable resistor r_1 r_2 balance can be restored (i.e galvanometer reading returned to zero) and whilst this is being done another pointer can be moved simultaneously and automatically to give the temperature – this is known as the null

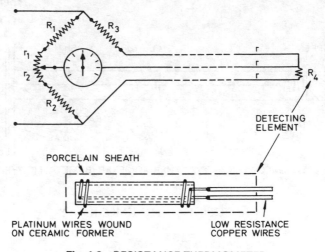

Fig. 1.2 RESISTANCE THERMOMETER

balance method Alternatively, the galvanometer can give the temperature reading directly, in this case no variable resistance r_1 r_2 would be required.

For the measurement of ambient temperature conditions the resistances, apart from the temperature measuring resistance, would have to be made of a metal whose resistance does not vary with temperature. A metal which nearly fulfills this requirement is constantan.

The term pyrometer is used for temperature measuring instruments operating above 500°C, the term thermometer being used for operation below 500°C.

Resistance thermometers can be exceedingly accurate. Platinum is the most suitable sensing wire element but copper and nickel wire are used in the range −100°C to 200°C and tungsten, molybdenum and tantalum are used to 1200°C, in protective atmospheres. The platinum element usually has a resistance of 100 ohms at 0°C in which case resistance of wires is limited to about 3 ohms. Use up to 600°C with twin wires is often acceptable with the three wire method used for higher accuracy; measurement is by Wheatstone, Kelvin or Mueller bridges or potentiometric methods.

THERMISTOR (THERMALLY SENSITIVE RESISTOR)

These devices are among a second class of resistance thermometer utilising elements made of semi-conducting material all of which have a characteristic of a resistance decrease with temperature increase. Included in this category are carbon resistors and doped germanium units (see Fig. 1.3).

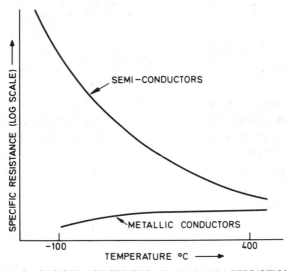

Fig. 1.3   RESISTANCE-TEMPERATURE CHARACTERISTICS

Thermistors are made of semi-conducting materials, they are manufactured by sintering (i.e. heating under pressure) powder mixtures of metallic oxides such as manganese, nickel, cobalt, copper, iron or uranium. The size and configuration can be controlled so that rods, beads, discs and washer shapes can be produced as desired. Figure 1.4 shows an ellipsoid of thermistor material, wires about 0·25 mm apart are firmly embedded in the material making good electrical contact. The whole assembly may be coated with glass to give strength and protection.

A washer shape of thermistor is particularly useful, it could be fitted over bolts or studs. Several washers could be used together with the terminals connected in series or parallel arrangements. Beads or rods of thermistor material are suitable for use as probes.

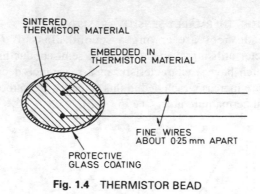

Fig. 1.4 THERMISTOR BEAD

Advantages of thermistors:

(1) Relatively small and compact, the bead arrangement shown could have a diameter up to 2·5 mm with a resistance up to about 100 megohms.

(2) Low specific heat, hence the thermistor does not take very much heat away.

(3) Physically strong and rugged.

(4) Relatively high temperature coefficient of resistance, it could be as high as ten times that of some metals.

(5) They can be used for extremely low temperature measurement with great accuracy.

The mathematical relationship for thermistors is given by:

$$\rho_\theta = \rho_0 e^\beta \ (1/\theta - 1/\theta_0)$$

where:

ρ_θ is the specific resistance at temperature θ

ρ_0 is the specific resistance at temperature θ_0

β is a constant which depends upon the material used in the construction ($\beta \simeq 4000$). The characteristic is shown in Fig 1.3.

The large negative temperature coefficient of resistance of thermistors may be explained by considering the number of electrons available for carrying current.

Few electrons are available at low temperatures but as the temperature increases the kinetic energy of the electrons increases and this enables them to move from inner tightly bound orbitals to the outer conduction bands of the atom. With more free electrons available to carry current the resistance to current flow reduces.

In metals, where there are many free electrons, increase in temperature leads to "traffic jams" of electrons hence the resistance to current flow increases with increase of temperature.

The very high sensitivity, small size, high energy and rapid response makes thermistors very useful detecting devices particularly for use in computers and scanner installations. Range is –100°C to 300°C but special compositions can extend this much further, as high as 1600°C. The small thermal mass can lead to self heating and coupled with high sensitivity and exponential characteristic means instability must be carefully watched. The thermistor merely replaces the resistance element on one limb of a bridge circuit in the measuring unit.

THERMOCOUPLE

Whenever the junctions formed of two dissimilar homogeneous materials are exposed to a temperature difference, an emf will be generated which is dependent on that temperature difference, also on the temperature level and the materials involved. This thermoelectric emf is called Seebeck effect (Seebeck discovered in 1821) and is an algebraic sum of two other effects discovered by Peltier and Thomson. The two materials, usually metals, form the thermocouple.

Figure 1.5 shows a thermocouple consisting of two wires, one iron, one constantan (*i.e.* a copper nickel alloy), with a millivoltmeter coupled to the copper wire. If the junctions A and B are maintained at the same temperature no current will flow around the circuit since the emfs in the circuit will be equal and opposite. If however A is heated to a higher temperature than B then current will flow since the emf at one junction will be greater than the opposing emf at the other junction.

A third wire can be introduced as shown in Fig. 1.5, where AB and AC form the couple wires. Providing the junctions B and C are maintained at the same temperature, the introduction of the third wire BC will not affect the emf generated. Hence A will be the hot junction and B with C will form the cold junction. Couple wires AB and AC shown as iron and constantan respectively can be made of various metals and alloys, choice depends upon temperature of operation, the wire BC would generally be long compared to the couple wires and could be made of copper. Figure 1.5 also shows the device in detail.

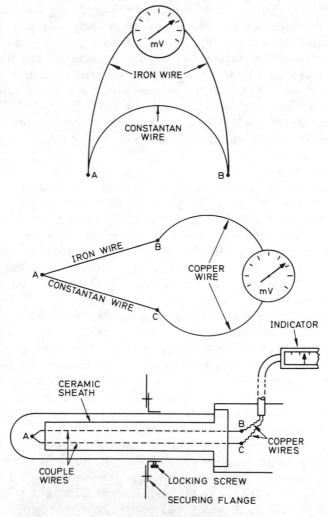

Fig. 1.5   THERMOCOUPLE

In a practical thermocouple system the cold junction B and C may be at a relatively high temperature due to the environment. This would mean that the temperature difference between the hot and cold junctions would be small and mV similarly. The indicator itself could then become the cold junction if the wires from terminals B and C to the indicators are of the same material, or material with similar characteristics, to the couple wires. The wires are then called compensating wires and the cold junction temperature would be reasonably constant if the indicator is within an air conditioned control room or immersed in a block of metal of large thermal capacity. Alternatively cold junction compensation signal by separate means or bi-metallic instrument components is arranged.

A copper (+) constantan (−) couple is used up to about 350°C, constantan being a 40% Ni 60% Cu alloy. Up to 850°C an iron-constantan couple is used with a chromel (90% Ni 10% Cr) and alumel (94% Ni 2% Al) couple up to 1200°C. Average emf is 0·05mV/°C which compares with about 18mV/°C for a thermistor. Platinum-platinum plus 10% rhodium couples have been used to 1400°C.

The emf generated is usually given by an expression of the form:

$$e = A + B\theta + C\theta^2 + D\theta^3 + \cdots$$

where θ is the temperature and A, B, C, D, are constants of diminishing order. Fig. 1.6 shows an electronic thermocouple with operational amplifier (A) giving 0.1 V/°C. Y is for calibration at cold junction temperature and X for other temperature calibration.

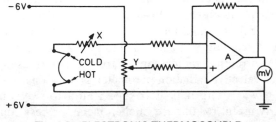

Fig. 1.6 ELECTRONIC THERMOCOUPLE

RADIATION PYROMETRY

When temperatures are above the practical range of thermocouples, or the "target" is not accessible, or an average temperature of a large surface is required then radiation pyrometers are used. Theory is generally based on black body radiation and the work of Stefan, Boltzmann and Planck with amended factors of emissivity to allow for variation from the ideal black body radiator. Types of radiation pyrometer are optical, radiation and two colour. The former will be considered.

OPTICAL PYROMETER

Referring to Fig. 1.7: S is the source and rays enter lamp box L after passing through the lens, aperture and absorption filter. The lamp is electric and current and voltage are measured at G. Rays leaving L pass through a red filter, lens and aperture to eye E.

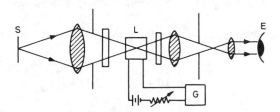

Fig. 1.7 OPTICAL PYROMETER

The device is often known as a disappearing filament unit. Both source and reference, the latter being the filament of a small vacuum lamp, are observed through the microscope. The power to the lamp is adjusted until the reference source just disappears into the main source. Power is calibrated to give a temperature reading directly. Correction factors apply for the filters used and the device is a selective radiation pyrometer as only a narrow band of radiation wavelength is utilised. Radiation pyrometry is particularly useful for furnace, molten metal, process control, etc. evaluation of temperature.

PHOTO-ELECTRIC PYROMETER

There are three types of photo-electric cells; the photo-emissive, the photo-conductive and the photo-voltaic; the latter is used here.

Incident light falls on p-type silicon layered on to n-type silicon backed with metallic strip. The emf generated is measured, after calibration, by a galvanometer or self balancing potentiometer connected across the p-type and the backing.

Such pyrometers are best suited to measuring small radiation sources and are stable and accurate with a very quick response time which makes them very suitable for distance remote control systems.

Note:

Many of the measuring devices for temperature considered previously, particularly electrical types, could also be classified as telemetering or transmitting units as the signal is readily conveyed over a considerable distance to a remote measuring, recording or display station. This applies in many cases to other such devices in the following chapters. The detecting element (sensor) is inherently a transducer in operation for many instrument units.

TEST EXAMPLES 1

1.   Describe, with the aid of simple sketches, three types of temperature measuring device. State how they are graduated and where they are used.

2.   Sketch and describe an electrical instrument used for reading temperature at a remote distance. State the usual temperature range and the materials used in construction.

3.   Give a reasoned explanation of how Wheatstone bridge networks are employed in circuits of electrical resistance thermometers, explosive gas sampling devices or any similar application. Sketch the circuit for such a device indicating the function of the Wheatstone bridge.

4.   Sketch and describe a temperature measuring system employing a thermocouple.

CHAPTER 2

PRESSURE MEASUREMENT

WATER MANOMETER

This instrument is used for measuring pressures of a low order such as fan pressures, etc. Fig. 2.1 illustrates a U tube water manometer, one limb of which is connected to the system whose pressure is to be measured, the other limb is open to the atmosphere. The pressure reading is the difference of the water levels read from the scale.

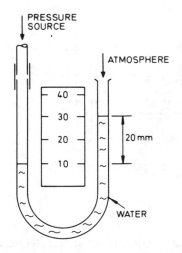

Fig. 2.1 WATER MANOMETER (U TUBE)

Note: 1 m³ of fresh water has a mass of 1 Mg and weighs 10^3 x 9·81 N.

Hence:

1 m of fresh water exerts a pressure of 10^3 x 9·81 N/m², 1 mm of fresh water exerts a pressure of 9·81 N/m², the reading indicated in Fig. 2.1 is therefore equivalent to 20 x 9·81 = 186·2 N/m², i.e. 0·1862 kN/m², above atmospheric pressure (1 bar = 10^5 Pa = 10^5 N/m² $\simeq$ 1 atm). A wide cistern manometer (well type) is used for lower pressure differentials (Fig. 2.2 as mercury manometer). A variation is the inclined tube manometer – the small bore limb is set at a small angle to the horizontal and the longer scale parallel to it gives even smaller differential pressure readings.

MERCURY MANOMETER

A mercury manometer of the well type is shown in Fig. 2.2.

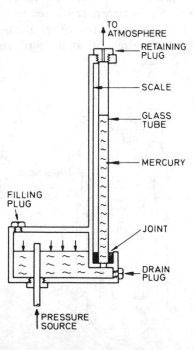

Fig. 2.2 MERCURY MANOMETER (WELL TYPE)

This instrument measures pressures of a higher order than that measured by the water manometer, such as scavenge or super-charge air pressure for IC engines. The uniform bore glass tube is small in diameter so that when mercury is displaced from the well into the tube, the fall in level of the mercury in the well is so small it can be neglected. Hence the pressure reading is indicted directly by the level of the mercury in the glass tube. The relative density of mercury is 13·6 hence 1 mm of mercury is equivalent to a pressure of 9·81 x 13·6 N/m², *i.e.* 134N/m² or 0·134kN/m². A special application is the vacuum gauge (kenotometer) which is a combined barometer and manometer with the scale on the right hand side calibrated in absolute pressure.

MERCURY BAROMETER

A sketch would be similar to Fig. 2.2 but with the top of the glass tube sealed at a vacuum and the pressure source would be the atmosphere.

If we assume the atmospheric pressure is supporting a column of mercury 760 mm in the tube, then:

$$\text{atmospheric pressure} = 760 \times 0·134$$
$$= 102 \text{ kN/m}^2$$
$$= 1·02 \times 10^5 \text{ N/m}^2$$
$$= 1·02 \text{ bar}$$

For an atmospheric pressure of 760 mm the equivalent water barometer would be 760 x 13·6, *i.e.* 10 336 mm or 10·336 m.

The value of the atmospheric pressure varies slightly with climatic conditions hence to ascertain true absolute pressures the barometer reading should be taken at the same time as a gauge pressure is taken.

If for example we wish to obtain the absolute pressure in a condenser and the readings were:

condenser gauge reading 742 mm

barometer gauge reading 762 mm

then, condenser pressure = (762 – 742) x 0·134

$$= 2·68 \text{ kN/m}^2$$

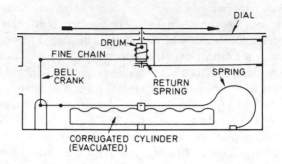

Fig. 2.3   ANEROID BAROMETER

ANEROID BAROMETER

The aneroid barometer is shown in Fig. 2.3. It consists of a corrugated cylinder (detecting element) made of phosphor bronze or other similar material, a steel spring, bell crank, pointer, dial and case (measuring element). The corrugated cylinder is completely evacuated hence the pressure of the atmosphere tends to collapse it. The centre of the corrugated cylinder deflects downwards if atmospheric pressure increases and the spring causes deflection upwards if atmospheric pressure decreases. Cylinder motion is transmitted to the instrument pointer.

PRESSURE GAUGE (BOURDON)

A pressure relay tube is the principal working component (detecting element). This tube which is semi-elliptical in cross section is connected to the pressure source. When the tube is subjected to a pressure increase it tends to unwind or straighten out and the motion is transmitted to the gauge pointer through the linkage, quadrant and gear (measuring element). If the tube is subjected to a pressure decrease it winds, or coils, up and the motion is again transmitted to the pointer. This gauge is therefore suitable for measuring pressures above or below atmospheric pressure. A diagrammatic sketch is shown in Fig. 2.4.

Materials used in the construction of the gauge are solid drawn phosphor bronze or stainless steel for the pressure relay tube. Bronze or stainless steel for the quadrant, gear and linkage. Case, brass or plastic.

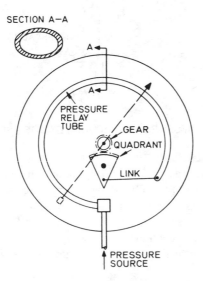

Fig. 2.4   PRESSURE (OR VACUUM) GAUGE – BOURDON

The Bourdon movement is frequently used in transducers and controllers to vary output signals in pneumatic or electrical form.

PRESSURE GAUGE (SCHAFFER)

This type utilises a strong flexible metal diaphragm (detecting element) which moves up as pressure increases. The device is shown in Fig. 2.5. Again this device can also be used as a transducer (pneumatic or electric) in telemetering or control with an output signal proportional to diaphragm movement. Similar remarks apply to most detecting (sensing) devices (as detailed in Chapter 6).

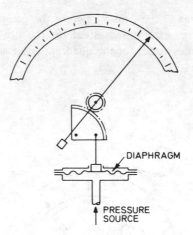

Fig. 2.5   PRESSURE (OR VACUUM) GAUGE – SCHAFFER

DIFFERENTIAL PRESSURE CELL (D/P CELL)

The sketch of Fig. 2.5 illustrates a single diaphragm subject to differential pressure. The d/p cell is often used in direct differential pressure recording as well as flow and level applications. The detecting element of the cell is a bellows, or diaphragm, whose mechanical movement is used to indicate or transduced to electrical or pneumatic signal output. Low pressure - slack diaphragm.

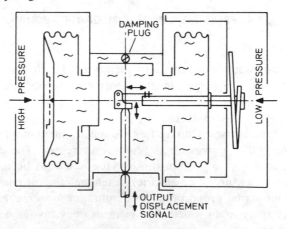

Fig. 2.6   DIFFERENTIAL PRESSURE CELL

As an alternative to a one membrane diaphragm a sealed capsule (twin membrane) can be inserted in the cell body and secured so that different pressures are applied at each side. The capsule can be filled with a constant viscosity fluid (for a fairly wide temperature range) which also damps oscillation. Silicone is such a fluid. Again mechanical movement of the capsule is proportional to differential pressure. Capsule stacks are also used.

Another type of d/p cell utilises two separate bellows. Such a design, often called a Barton cell, is shown in Fig. 2.6. Refer now to this sketch.

Pressure increase on the high pressure side displaces liquid (via an adjustable damping plug) and expands the right hand low pressure bellows. This bellows is connected to the horizontal spindle, one end of which has a flexure strip and the other end is fastened to flat plate springs. Equilibrium exists when spring force equals differential pressure. Mechanical travel is via the flexure strip, lower seal diaphragm and vertical spindles to indicator or transducer. A bi-metallic strip adjusts bellows fluid capacity to allow for volumetric expansion. This device is shown dotted within the left hand bellows. The bellows are often made of welded stainless steel discs and inner faces butt if excess pressure is applied, so protecting the bellows.

PIEZOELECTRIC DETECTING ELEMENT (SENSOR)

With *certain* solid crystals having an asymmetrical electric charge distribution, any deformation of the crystal produces equal external unlike electric charges on the opposite faces of the crystal. This crystal is known as the piezoelectric effect.

Deformation of the crystal can be caused by pressure and the charges produced can be measured by means of electrodes attached to the opposite surfaces of the crystal.

Crystal materials can be naturally occurring or man made. Quartz (*i.e.* SiO_2) is a material that can be used in temperature environment up to 550°C, it is stable mechanically and thermally. Barium titanate, a ceramic produced commercially, is also used but tourmaline is principally used because of its good electrical properties. The output voltage from tourmaline crystal, acting as a detecting element (sensor) and transducer, is a linear function of the pressure applied and the charge sensitivity is approximately 2 x 10^{-10} coulomb/bar, it can be used for pressures varying from 1·03 to 800 bar. The only drawbacks are (1) its sensitivity to temperature change (*i.e.* its pyroelectric effect) hence ideally it should only be used under controlled temperature conditions (2) it is more suited to pressure variations than to static pressure.

Figure 2.7 shows crystals in series and parallel. The series arrangement gives a higher output voltage for the same pressure applied whereas the parallel arrangement reduces output impedance and hence is a more stable system. Figure 2.8 is a commercial piezoelectric sensor using tourmaline crystals in parallel, the sensor is small and is coated with insulating material to help minimise the effect of temperature variation in the medium whose pressure is being measured.

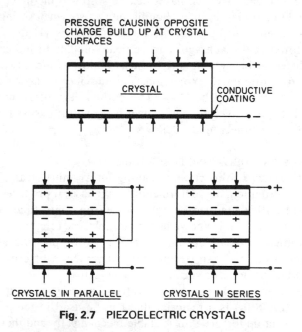

Fig. 2.7 PIEZOELECTRIC CRYSTALS

Pyroelectricity, like piezoelectricity, makes use of the spontaneous polarisation of materials such as ferro-electrics. Temperature variations causing changes in polarisation can be utilised in very accurate calorimetric recordings.

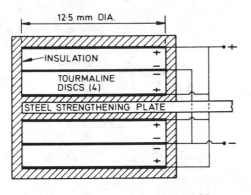

Fig. 2.8 PIEZOELECTRIC SENSOR

STRAIN GAUGE

A strain gauge is fundamentally a resistive wire of about 0·01 mm diameter subject to strain by pressure (force or acceleration) with electrical resistance change proportional to strain. *Bonded* elements of the wire wound type are either flat grid or helical wound on a former. The wire is fixed to a backing material such as paper, resin or plastic which is glued to the surface under test, wires are soldered or spot welded. One alternative is the foil type where the grid is etched from thin metal foil using printed circuit techniques. Another alternative is a *p* or *n* doped silicon semi-conductor which can be very small and is extremely sensitive.

The *unbonded* strain gauge is essentially a pressure sensor and a typical design is shown in Fig. 2.9.

The detecting (sensing) element can be fastened directly to the diaphragm as shown, or alternatively, a central force rod can transmit diaphragm movement to the detecting (sensing) element consisting of plate springs with posts on the periphery upon which the strain gauge wire is wound.

In all strain gauges, to minimise resistance change due to temperature effects, it is usual to employ materials with a low temperature coefficient of resistance for the wire, such as constantan. Alternatively a second compensating wire loop can be incorporated.

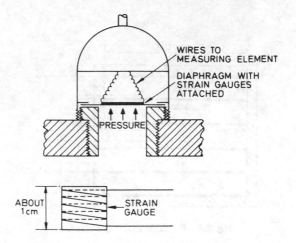

Fig. 2.9   STRAIN GAUGE PRESSURE SENSOR

The measuring element for strain gauges is generally a Wheatstone bridge circuit with temperature compensating resistance and strain gauge resistance arranged as two, of the four, resistances and a central galvanometer and constant dc voltage source. Null balance methods can also be used, or for high sensitivity the bridge may be fed by ac voltage and the galvanometer replaced by a transistor amplifier and detector. The measuring unit is effectively a transducer, *i.e.* displacement-electrical..

TEST EXAMPLES 2

1. Sketch and briefly describe three types of pressure measuring device.

2. Sketch and describe a Bourdon type of pressure gauge. State the materials used in construction. Discuss briefly how the Bourdon movement can be utilised for telemetering devices.

3. Describe, with the aid of a sketch, any type of differential pressure cell. Detail three applications in instrumentation of the use of the d/p cell.

4. Explain the operation of a foil strain gauge and describe a bridge method to achieve an output voltage proportional to the strain.

CHAPTER 3

LEVEL MEASUREMENT

Liquid level sensors are usually classified under two headings, i.e. direct methods and inferential methods.

DIRECT METHODS

FLOAT OPERATED

The float is generally a hollow cylinder or ball working on direct action or displacement principles. Level variation is transmitted by chain or wire and pulley or torque tube (usually with counterweights fitted) to the indicator. High or low level alarm contacts are easily arranged. Pulley movement can also be arranged to operate a contact arm over an electrical resistance so varying current or voltage to indicator or receiver.

SIGHT GLASSES

Various types are in use dependent on working conditions. The simple boiler water glass gauge with toughened glass and the plate type of water gauge for high pressures are typical.

PROBE ELEMENTS

Floatless types of level sensors can be arranged where the liquid is a conductor. Sensing electrodes, rods or discs, vary electrical circuits when they are in contact with liquids. A typical example is detection of the fluid level in a tank by capacitive techniques (see Fig. 3.1). This technique is used in oil-water interface detection on oily-water separators - see Chapter 13.

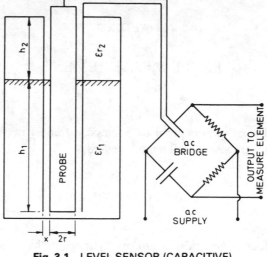

Fig. 3.1   LEVEL SENSOR (CAPACITIVE)

Detection of level is measured by variation of capacitance which is accomplished by alteration in dielectric strength. One plate of the capacitor is a probe, possibly made of stainless steel, the other is the shell and both are connected to an ac bridge which is supplied with high frequency low voltage alternating current. As the interface moves the dielectric strength (relative permittivity ε_r) alters.

An approximate expression for the capacitance in an ac bridge circuit (Fig. 3.1) is:

$$C = \frac{\varepsilon_0[\varepsilon_{r1}h_1 + \varepsilon_{r2}h_2]}{2l_n[1 + x/r]}$$

and since ε_{r2} is approximately unity for air ($\varepsilon_{r1} > 1$)

$$C = \frac{\varepsilon_0[\varepsilon_{r1}h_1 + h_2]}{2l_n[1 + x/r]}$$

where ε_0 is the free space permittivity, ε_r relative permittivity
" h_1 " "   head of fluid being measured
" h_2 " "   distance from fluid surface to tank top
" $2r$ " "   diameter of the probe
" x " "   separation between the circular plates
" l_n " "   natural logarithm.

INFERENTIAL METHODS

PRESSURE ELEMENTS
   The static-pressure method is extensively used.

$$p = \rho g h$$

p is pressure, ρ density of fluid, g gravitational acceleration, h fluid head. Any of the sensing and measuring devices described in Chapter 2 are therefore applicable to level measurement. In particular a pressure gauge, calibrated in height units, is probably the most simple indicator. Bellows or diaphragms are used with the pressure bulb located inside or outside the vessel with suitable correction factors applied for correct datum and density of fluid. The differential pressure cell is often used in level measurement. Telemetering for remote reading is readily applied, pneumatically or electrically, to the displacement movement of the level (pressure) sensor.

MANOMETER TYPES
   The static pressure equation is applicable and devices have been described in Chapter 2. Three types of remote level indicator are now described.
   For the Electroflo electrical type as sketched in Fig. 3.2 the difference in level h is directly proportional to the difference between the level of the liquid in the tank and a datum. The transducer element contains resistances immersed in mercury to form an electrical circuit with transmission to remote indicator receivers.

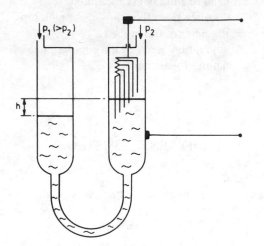

Fig. 3.2   LEVEL SENSOR/TRANSDUCER (RESISTIVE)

Figure 3.3 is a diagrammatic arrangement of the Igema remote water level indicator. The lower portion of the U tube contains a (red) coloured indicating fluid which does not mix with water and has a density greater than that of water.

The equilibrium condition for the gauge is $H = h + \rho x$ where ρ is the relative density of the indicating fluid. H, h and x are variables.

If the water level in the boiler falls, h will be reduced, x will be increased and H must therefore be increased. The level of the water in the condenser reservoir being maintained by condensing steam.

If the water level in the boiler rises, h will be increased, x will be reduced and H must therefore be reduced. Water will therefore flow over the weir in the condenser reservoir in order to maintain the level constant.

A strip light is fitted behind the gauge which increases the brightness of the (red) indicating fluid, which enables the operator to observe at a glance, from a considerable distance, whether the gauge is full or empty.

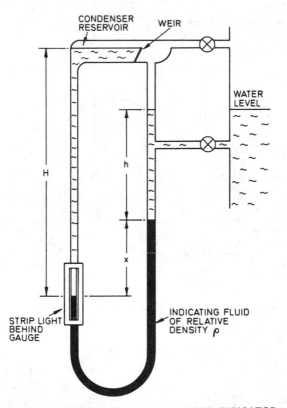

Fig. 3.3 IGEMA REMOTE WATER LEVEL INDICATOR

Figure 3.4 is another type of remote water level indicator. In this case the operating fluid is the boiler water itself. The operation of the gauge is as follows:

If we consider a falling water level in the boiler, the pressure difference across the diaphragm h will increase, causing the diaphragm to deflect downwards. This motion of the diaphragm is transmitted by means of a linkage arrangement (see insert) to the shutter which in turn moves down pivoting about its hinge, causing an increase in the amount of (red) colour and a decrease in the amount of (blue) colour seen at the glass gauge.

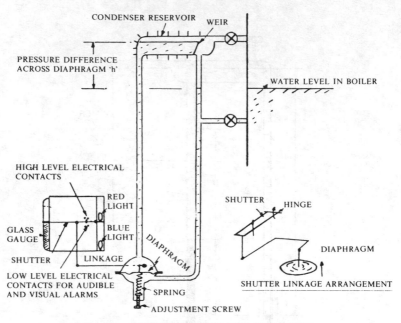

Fig. 3.4   REMOTE WATER LEVEL INDICATOR

It will be clearly understood that if the water level now rises then the (red) will be reduced and the (blue) increased.

Separating the (blue) and (red) colours, which are distinctive and can clearly be seen from a considerable distance, is a loose fitting black band which moves with the shutter, giving a distinct separation of the two colours.

An adjustment screw and spring are provided to enable the difference in diaphragm load to be adjusted. Hence correct positioning of the shutter and band in relation to the reading of a glass water gauge fitted directly to the boiler is possible.

Both devices shown in Figs 3.3 and 3.4 are also capable of being observed by closed circuit television systems to extend the distance between transmission and reception. Alternatively telemetering by pneumatic or electrical transmitters is readily arranged. A d/p cell would be very suitable for the former and any displacement transducer, such as a differential transformer, could be used for the latter.

PURGE SYSTEMS

For small air flow rate, about one bubble per second, a pressure equal to that in the dip tube will be applied to the indicator as shown in Fig. 3.5. This simple bubbler device is an arrangement that is similar to the well known **pneumercator** used for determining depths of water and oil in tanks. Air supply to the open ended pipe in the tank will have a pressure which is directly proportional to the depth of liquid in the tank.

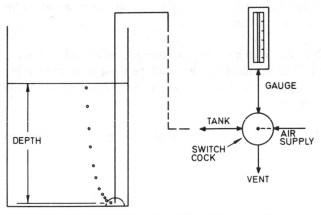

Fig. 3.5   PNEUMERCATOR LEVEL INDICATOR

ULTRASONIC AND NUCLEONIC DEVICES

An *ultrasonic* transmitter and receiver (utilising piezo-electric crystals) are located above the tank so that two echoes are received, one from the liquid surface and one from the tank bottom. The time separation between the two signal echoes is proportional to liquid depth - which can be displayed and measured by a cathode ray oscilloscope (CRO).

Nucleonic units have a shielded radioactive strip source from which gamma radiation is picked up by a detector on the opposite side of the storage vessel.

TEST EXAMPLES 3

1. Sketch and describe a distant-reading boiler water level gauge. Explain its principle of operation.
Give two ways in which the gauge would give incorrect readings.
State the routine maintenance necessary to ensure maximum reliability at all times.

2. Sketch and describe any type of level sensor. Discuss the modifications necessary to utilise the device as a transmitter in a telemetering system.

3. Describe, with a sketch, an instrument used to indicate the level of liquid in a tank. Can the result be used to determine pressure above a given datum and, if so, state the other variables likely to be involved.

4. Describe the arrangement of an air purge tank contents system.

CHAPTER 4

FLOW MEASUREMENT

Many of the techniques utilised in flow measurement employ principles also used in pressure and level measurements.

BASIC PRINCIPLES

Flowmeters are generally divided into two fundamental types, *i.e.* quantity meters and rate of flow meters, use of the word flowmeter generally implies the latter.

QUANTITY METERS

These devices measure the *quantity* of fluid that has passed a certain point. *No time is involved.* Types are usually classified as *positive* or *semi-positive*. A typical positive type utilises the flow to drive a reciprocating piston and a counter is attached. The meter acts like a conventional engine with fluid pressure supplying motive power. Stroke length and cylinder dimensions fix the quantity delivered per cycle. Semi-positive types are usually rotary. A form of gear pump, or eccentrically constrained rotor, can be used which is driven by the fluid. Quantity is measured by number of rotations (cycles) and fluid per cycle.

RATE OF FLOW METERS

Measure the velocity of fluid passing a certain point at a given instant. From this rate of flow (quantity per unit time) can be determined from velocity multiplied by area of passage. They are therefore classified as *inferential, i.e.* volume inferred from velocity. There are two fundamental components of the rate of flow meter. The *primary* element is that portion of the instrument which coverts the quantity being measured into a variable to operate the secondary element, for example, orifice and pressure tappings from a venturi. The *secondary* element measures the variable created by the primary element, for example, a differential pressure cell.

INTEGRATORS

Quantity meters are more expensive and less suited to deal with large fluid quantities than rate of flow meters. Rate of flow meters are often used as quantity meters by fitting an integrator. As a simple example consider variable flow. The rate of flow can be measured at set time intervals and a graph plotted, the area of which gives quantity over the time period required. In practice this is performed mechanically or electrically by an integrator. One type is based on the planimeter principle and another type (escapement) utilises mid-ordinate techniques from a heart-shaped cam drive. Flat faced cam drive, or worm and wheel designs, can be used with a turbine wheel or helix type of primary element having a counting mechanism secondary element, incorporating the integrator to interpret quantity. The integrator is often included within the receiving unit of telemetering systems and a typical device is illustrated and described in Chapter 6. Integration is readily performed electrically by use of a conventional watt-hour meter. It should be understood that integration is a general instrumentation operation whose use is not restricted only to flowmeters.

SQUARE ROOT EXTRACTION

When inferential devices are used, with velocity sensors utilising differential pressure techniques, the velocity is not directly proportional to pressure difference, or head. Velocity is related to the square root of pressure, or head, i.e. a curve of flow rate plotted against pressure, or head, is of parabolic form.

This means that if a pressure difference is used in a sensor device connected to a manometer, or pointer through a linear mechanism, the rate of flow scale on the manometer or traversed by the pointer would have to be a square root function. The scale divisions would increase in square root increments for equal increments of flow rate. From the aspect of display and continuous recording of flow rate the square root characteristic is not an embarrassment. If however the differential pressure has to be used in a control system the square root is usually extracted to give a signal which is directly proportional to the fluid flow rate. Square root extraction is described later in this chapter under differential pressure inferential devices. It is difficult to integrate readings of a system when the square root extraction is not applied.

INFERENTIAL-ROTATIONAL

Basically these can be considered as mechanical or electrical.

MECHANICAL TYPE FLOWMETER

Designs are usually of the "turbine wheel" type with speed of rotation directly proportional to linear flow velocity and, with area of passage fixed, the volume rate is inferred. A wheel, fan or helix is inserted in the pipe or duct, mounted vertically or horizontally, gear trains are used to interpret the movement. This is the principle of vane anemometers.

Rotormeter

The measuring principle is illustrated in Fig. 4.1 where the meter, of the duplex rotor positive displacement type, is shown in three specific positions. This rotary flowmeter operates on the displacement principle, the measuring system consisting of a casing with two rotors. Bearing bushes are provided on either side of each rotor so that the rotor runs clear of the casing and spindle in a radial direction. The rotor is located in an axial direction by means of end bearings so that it runs clear of the bearing plates. Both the end bearings and the spindle are fixed to the shaft whilst the bearing bushes are fixed to the rotor and rotate relative to the end bearings and the shaft. Each rotor carries a gear wheel at the rear through which the rotors are coupled together. The rotary movement of the rotors is transmitted to the outgoing shaft of the meter through a gearwheel fitted at the front of one of the rotor. Consider Fig. 4.1:

In position "A", the left-hand rotor is fully relieved from load whilst the liquid pressure acts on one side of the right-hand rotor

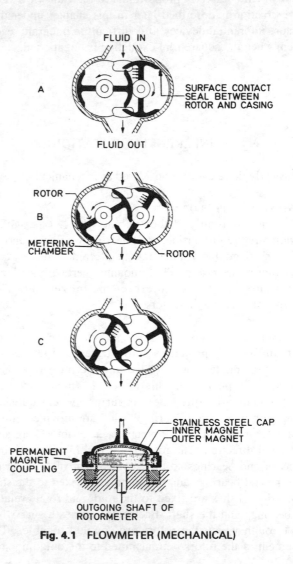

Fig. 4.1 FLOWMETER (MECHANICAL)

causing this rotor to rotate in clockwise direction. Since both rotors are coupled together through gears, the left-hand rotor will rotate in an anti-clockwise direction.

In position "B", the liquid in the displacement chamber is pressed by the right-hand rotor to the outlet.

In position "C", the right-hand rotor is entirely relieved from load whilst the liquid pressure now acts on one side of the left-hand rotor so that it takes over the task of the right-hand rotor.

To reduce leakage losses to a minimum, it is essential to provide for effective sealing between inlet and outlet. To this end, the rotors seal off against the casing by surface contact.

In the rotormeter, the motion of the rotors is transmitted to the external parts by the attraction between two permanent magnets, an inner and outer magnet. The maximum torque transmitted by this system is 0·4 N m. This arrangement offers the following practical advantages;

(1) Perfectly leakproof transmission, which means that it is impossible for corrosive or hazardous liquids to leave the meter.

(2) Protection of attached parts or instruments. If the mechanism of external parts or instruments should be blocked for whatever reason, the permanent magnetic coupling will slip, thus avoiding any damage to the instruments.

Although the torque to be transmitted by the magnets may vary with the external parts or the type of instrument used, it will hardly ever exceed a value of 0·05 N m with the meter running at constant speed. Hence, at ambient temperature the magnetic coupling has an 8-fold safety margin which will, in most instances, be amply sufficient to take accelerations and decelerations (which are the most frequent operating conditions) without slipping of the coupling.

Although the torque transmitted by the magnets decreases with rising temperature, experience has shown that even with a liquid temperature of up to 250°C the available torque is still amply sufficient.

ELECTRICAL TYPE FLOWMETER

One type utilises rotating vanes with a small magnet attached to one vane and a coil in the pipe wall. The electrical impulse can be counted on a digital tachometer calibrated in flow rate. The design now described has no moving parts.

Electro-magnetic flowmeter

This type is shown in Fig. 4.2.

The principle utilised is that of a moving conductor (the liquid) in a magnetic field generating a potential difference. In the simple arrangement shown the two electro-magnets are supplied with current (ac is preferred to dc to reduce polarisation of the dielectric). There are two sensor electrodes. If B is the flux density of the field, v velocity of flow, d pipe diameter, then in suitable units the emf generated at any instant is given by:

$$e = Bvd$$

For constant B and d, e is directly proportional to v.

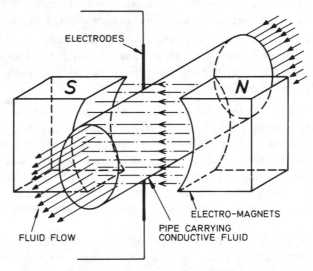

ELECTRODES

S   N

ELECTRO-MAGNETS

PIPE CARRYING
CONDUCTIVE FLUID

FLUID FLOW

Fig. 4.2 FLOWMETER (ELECTRICAL)

ROTAMETER

This type does not strictly fit into the classifications given but a brief description is appropriate at this stage. A sketch is given in Fig. 4.3.

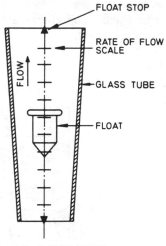

Fig. 4.3 ROTAMETER

This is a variable area meter. A long taper tube is graduated on its vertical axis. The float moves freely in the tube and by an arrangement of shaped flutes in the float it slowly rotates. As flow rate increases the float rises in the tube, so that the annular area increases, which means that the differential pressure across the tube is at a constant value. The float can be arranged with a magnet attachment and a follower magnet outside will transmit motion to a pointer via linkage if required.

INFERENTIAL-DIFFERENTIAL PRESSURE

PRIMARY ELEMENTS

The orifice and the venturi will be described and these are mainly used although flow nozzles and special inserts, such as a Dall tube, are also employed. Fig. 4.4 shows both the orifice plate and venturi sensors using energy conversion to produce a pressure difference which can be utilised by the secondary element to provide a signal for direct reading, telemetering or control.

Using the venturi flow sensor as an example the theory involved is as follows:

Assuming unit mass, and energy at points 1 and 2 being the

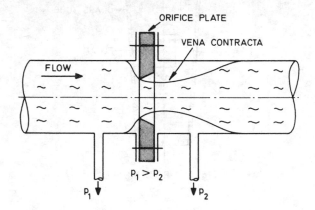

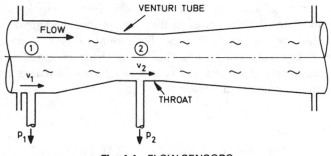

Fig. 4.4 FLOW SENSORS

same, i.e. neglecting friction and shock losses as small, then from Bernoulli for incompressible flow of fluid of density ρ:

$$KE \text{ at } 1 \ + \ PE \text{ at } 1 \ = \ KE \text{ at } 2 \ + \ PE \text{ at } 2$$

$$\tfrac{1}{2}v_1{}^2 \ + p_1/\rho \ = \ \tfrac{1}{2}v_2{}^2 \ + p_2/\rho$$

where KE is kinetic and PE is pressure energy.

The equation for continuity of flow for area A is;

$$v_1A_1 = v_2A_2$$

By substituting for v_2 from (b) in (a) and using mass flow rate $\dot{m}$ as equal to ρv_1A_1 it can be shown that

$$\dot{m} = k\sqrt{p}$$

where p is the pressure difference $(p_1 - p_2)$ and k is a meter constant in terms of areas and density which includes a discharge coefficient factor. Frictional losses are greater for orifice than for venturi meters. Seal pots protect sensor leads.

SECONDARY ELEMENTS

Any differential pressure device can be used as a secondary element including the manometer, diaphragm and d/p cell as described in Chapter 2. The measure scale will be non-linear for direct recorders due to the square root relation and telemetering, control and integration will be generally unsatisfactory unless a correcting unit is fitted. When manometers are used various compensations can be used. The simple manometer can utilise a curved measuring limb and the well type manometer can be arranged with a shaped chamber or may include a parallel tube and a shaped displacer. Other direct measuring devices utilise a cam incorporated in the mechanism, an example is the ring balance. In the electrical resistive sensor described in Chapter 3 (Fig. 3.2), when used for flow measurement, the electrode tips immersed in mercury are arranged in a parabolic curve with each other which gives the compensation. Three flow sensor square root extraction devices are now considered in more detail.

SQUARE ROOT EXTRACTORS

Figure 4.5 is one type of square root extractor using a parabola shaped bell which can be connected through linkage to mechanical, pneumatic or electrical display and control systems.

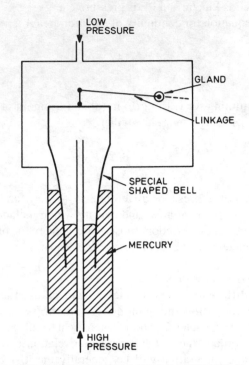

Fig. 4.5   SQUARE ROOT EXTRACTOR (MECHANICAL)

The differential pressure is applied with the high pressure inside and the low pressure outside the bell.

With its changing cross sectional area and buoyancy, due to change in differential pressure, the bell movement is made to be directly proportional to the square root of the differential pressure. Hence the bell movement is directly proportional to the fluid flow being sensed by the orifice plate or venturi sensor.

This device is often called a Ledoux Bell. For air flow measurement, instead of the bell being shaped. a shaped displacer is arranged in a separate chamber. The displacer is connected to rise with the bell. Such a device is used for measuring steam and air flow, and for controlling steam/air flow ratio, the characteristic is linear.

Figure 4.6 is a diagrammatic sketch of another type of square root extractor utilising the pneumatic flapper nozzle position balance principle (as described in Chapter 6).

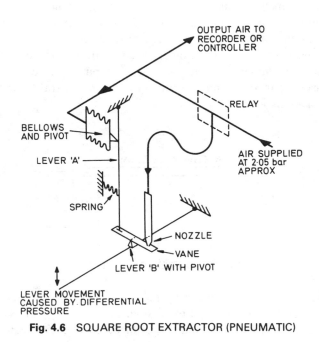

Fig. 4.6 SQUARE ROOT EXTRACTOR (PNEUMATIC)

Refer to Fig. 4.6:

The differential pressure from a flow sensor acts on a horizontal lever B, this effects the amount of air escaping from the nozzle and hence the pressure in the bellows. Pressure alteration in the bellows causes movement of the vertical lever A. The very small relative motion between levers A and B provides square root extraction. The output air signal is directed to the measure element (recorder) or controller.

Fig. 4.7 shows a square root extraction technique using electrical force balance (as described in Chapter 6). The differential pressure, in electrical signal form represented as variable x input, is applied to the left hand side. With the force balance beam in equilibrium the output signal is variable $\sqrt{x}$. Variation in input signal, causing unbalance, can be arranged to be re-balanced by suitable adjustment to output signal. This can be done in various ways, one method could be to connect the right hand of the beam to a differential capacitor in the output circuit (see Fig. 6.6). This is a closed loop but without electrical connection between input detected signal and output measured signal. The force balance system is a simple analogue computer - output can also be easily arranged for squaring, summing, etc, by suitable wiring combinations.

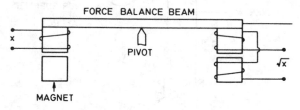

Fig. 4.7   SQUARE ROOT EXTRACTOR (ELECTRICAL)

Figure 4.8 shows two arrangements of flow sensor/transducer units, each with square root extraction. For the pneumatic system the valves marked A, B and C would be used in the sequence, open B, close A and C, when taking the differential pressure transmitter out of operation. This valve sequence would also apply to the electric system.

With steam flow measurement, the pressure tappings from the flow sensor are led to cooling reservoirs wherein the steam condenses. Water only then acts on the square root eliminator or recorder thus preventing damage.

The pneumatic system would utilise square root extraction as shown in Fig. 4.6 (the technique shown in Fig. 4.5 could also be used).

The electrical system can utilise mechanical movement (from a device such as Fig. 4.5) as shown to give square root extraction. Variation of the position of the soft iron core of the inductor unit

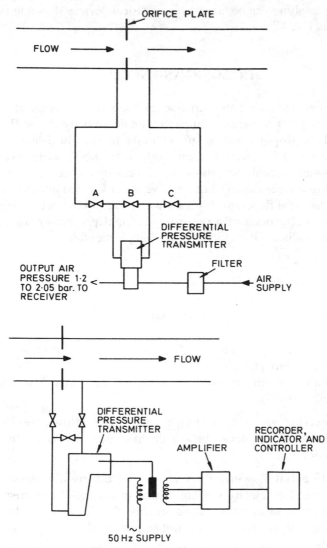

Fig. 4.8 FLOW SENSORS/TRANSDUCERS

govern output signal to the amplifier. Alternatively direct signals to the amplifier can be taken from electrical devices shown in Fig. 3.2 and Fig. 4.7.

ULTRASONIC AND NUCLEONIC

Ultrasonic flowmeters utilise an oscillator transmitter and receiver timer to measure apparent sound velocity in the fluid, which is proportional to the velocity of the fluid, across a diametral path inclined at an angle with the flow direction. Calibration is required to allow for the velocity profile.

Nuclear types employ a radioactive tracer injected into the fluid to measure the flow time, for a known volume, between two points by two detectors and a recorder. Normal protection against ionising radiation is required for the sample measure.

TEST EXAMPLES 4

1. Sketch the apparatus and describe the method of measuring steam flow through a pipe.
Explain why results obtained using this apparatus may differ from theoretical results.

2. Describe, utilising a sketch, any type of flowmeter. For inferential types of device how is the measured variable related to flow rate?

3. Inferential flowmeter devices using differential pressure techniques for velocity sensing exhibit non-linear characteristics. Discuss these characteristics and describe a device to restore linear characteristic to measure or control output signal.

4. Sketch an orifice plate installation showing the flow pattern and pressure variation. To what is the flow rate proportional?

CHAPTER 5

OTHER MEASUREMENTS

From an engineering viewpoint the variables discussed in the preceding chapters are the most significant. However, other variables have a particular specialist interest and a selected number are now presented with the appropriate instrumentation measurement.

ELECTRICAL

Such instrumentation covers a wide field, generally well described in electrical text books, and on this basis is not considered in this text unless specifically necessary. The measurement of capacitance, current, voltage, resistance, frequency, phase, energy, etc. are all very important in instrumentation. The reader will be familiar with such as moving iron and moving coil meters, etc. The bridge (ac or dc) and potentiometer are frequently used in telemetering and control components. Electronic measuring devices are being increasingly used and rectifiers, amplifiers, oscillators, etc. are an integral part of telemetering and control systems.

SPEED-TACHOGENERATOR

The dc tachogenerator is a small precision generator driven by the shaft whose rotational speed is required. Output voltage is directly proportional to speed. The tacho is best geared to run at maximum speed so giving maximum output signal and improved signal to noise ratio.

When used as a *tachometer* in measuring systems the output voltage from the tachogenerator is measured on a conventional voltmeter calibrated in terms of rotational speed (an analogue device).

Mechanical type tachometers are based on centrifugal action, linked to produce lateral travel.

A conventional ac generator for use as a tachogenerator or tachometer is generally not satisfactory because frequency (and phase) as well as voltage amplitude are proportional to speed of rotation. In this case a "drag-cup" device is used. The rotor is a thin aluminium cup rotating around a fixed iron core. The stator is wound with two coils at right angles, one ac supply the other ac output. With the cup stationary there will be no output as the windings are at right angles. Induced emf with cup rotation, due to cutting of flux of supply winding, links with the output winding so giving a signal proportional to rotational speed; frequency and phase being that of input signal. The device can be used for rate of change detection. With dc supply at constant speed no emf is induced in the output coils but angular acceleration or deceleration induces a voltage proportional to this change in the output coils. A "velocity voltage" applied to a differentiation (rate of change) circuit (CR series) will give a voltage across the resistor which is approximately proportional to acceleration (especially with a small time constant). Alternatively as force is proportional to acceleration a simple spring accelerometer can be used. A digital tachometer (counter) is shown in Fig. 5.1.

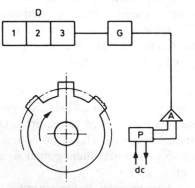

Fig. 5.1   DIGITAL TACHOMETER

As the (ferrous) toothed wheel rotates each tooth alters the air gap and flux in a pick up coil (P) whose output pulses are amplified (A). Pulses pass through a timing gate (G), say one second opening period, and are counted on a digital counter (D) which scales (related to teeth number per revolution) and displays as revolutions per second. Alternatively rev/min readings can be arranged with different gate, or scale, settings.

TORQUE-POWER

Indicated power can be measured by a conventional mechanical indicator although modern practice is tending towards oscilloscope display with integration for power. Shaft power of engines is measured by a torsionmeter in conjunction with a tachometer (power proportional to product of torque and rotational speed). Specific fuel consumption is readily achieved from these readings with a flowmeter calibration of fuel consumption. Various types of torsionmeter are available but those giving a continuous reading are usually of the electrical type. One design in common use is based on differential transformer operation (Chapter 6) which is illustrated in Fig. G of the specimen questions at the end of the book. Another design is based on magnetic stress sensitivity and is termed a *torque inductor—torductor*—and is now described.

The torductor is, as the name implies, a *torque inductor,* it is a stress transducer that is eminently suited to the measuring of torque in rotating shafts. It gives a high power output and requires no slip rings or other shaft attachments since it operates without any contact. Figure 5.2 shows a ring torductor. It consists of one primary ring which carries four poles, marked N,S, that is supplied with (50 Hz) alternating current. Two outer secondary rings have four poles each, arranged at 45° to the primaries, all of which are connected in series with mutually reversed windings.

No contact exists between the poles and the shaft, there being a 2 to 3 mm air gap provided to ensure this.

When no torque is applied to the shaft there are no stresses in the shaft and the magnetic fields between NS poles induced in the shaft will be symmetrical, the equipotential lines are then situated symmetrically under the secondary poles S_1, S_2, as shown and secondary flux and voltage will then be zero.

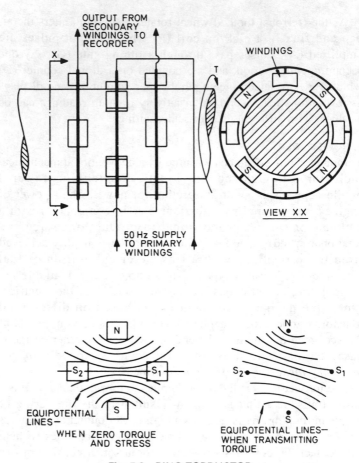

Fig. 5.2   RING TORDUCTOR

When a torque is being transmitted the equipotential lines form an asymmetrical pattern, as shown, due to the mutually perpendicular unlike stresses, acting at 45° to the shaft axis, causing increased permeability in one direction and decreased permeability at 90° in the other direction. This causes the S_1 pole to become magnetically slightly positive and the S_2 pole slightly negative.

The output from the secondaries of the ring torductor is of the order of a few milliwatts, which is large enough to be used without any amplification. If this signal is now married to a speed signal, from a tachogenerator perhaps, then the power being developed could be displayed directly on to one dial.

VISCOMETER

Newton investigated the viscosity of fluids and postulated, for most fluids under prescribed conditions, that flow rate is proportional to applied stress, more exactly that applied shear stress is proportional to velocity gradient.

$$\frac{F}{A} = \eta \frac{dv}{dx}$$

where η is a constant called the coefficient of viscosity. Applying this equation to the case of flow through a small bore tube of radius r and length l gives:

$$\eta = \frac{\pi p r^4}{8l\dot{V}}$$

For a constant flow rate ($\dot{V}$):

$$\eta = p \times \text{a constant}$$

where p is differential pressure.

Figure 5.3 shows the operational arrangement of the sensor element of a viscometer (see Fig. 12.12). A small gear pump driven at constant speed, by an electric motor through a reduction gear, forces a *constant* fluid quantity from the housing through a small bore tube (capillary). Fluid flows through the capillary without turbulence, *i.e.* streamline (laminar) flow prevails and pressure differential is proportional to viscosity of fluid. The pressure differential can be measured by any of the means previously described. The device, within a control system, is described in Chapter 13 (Fig. 13.19).

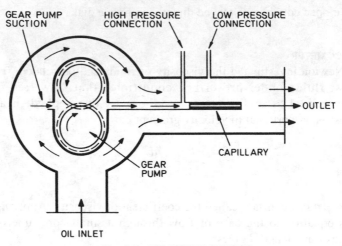

Fig. 5.3 VISCOSITY SENSOR

PHOTO-ELECTRIC CELLS

Photo-conductive cells are constructed with a thin layer of semi-conductor material and their resistance varies with the incident light energy. They are used as sensors in many situations such as oil-water content, smoke density, oil mist, flame indicator, etc. detection as described later in this chapter.

Photo-emissive cells rely on the light energy providing energy to release electrons from a metallic cathode.

If visible light, which is radiation and hence energy, falls upon certain alkali metals — such as caesium — electrons will be emitted from the surface of the metal. Metals in general exhibit this characteristic but for most materials, the light required has a threshold wavelength in the ultra violet region so that visible light does not cause electron emission.

Light energy comes in packages called *photons* and the energy of the photons is used in doing work to remove the electrons and to give the electrons kinetic energy after escape from the metal.

Figure 5.4 shows a simple photocell, visible light falls on the metal cathode from which electrons are emitted, they collect at

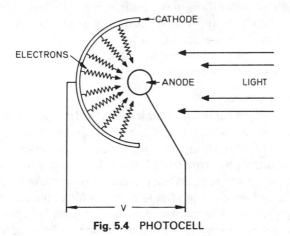

Fig. 5.4 PHOTOCELL

the anode and in this way create a potential *V* which can then be amplified and used for alarm and control, etc.

In the vacuum cell all current is carried by photo electrons to the positive anode. In the gas filled cell emitted electrons ionise the gas, producing further electrons, so giving amplification. Secondary-emission (photo-multiplier) cells utilise a series of increasingly positive anodes and give high amplification.

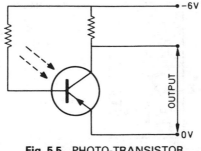

Fig. 5.5 PHOTO-TRANSISTOR

Photo-transistors exhibit similar characteristics and small size and high amplification make their use particularly attractive especially when applied to counting systems, *i.e.* digital tachometry. This device is shown in Fig. 5.5. Optical focus incident light on to the base increases the base current, hence collector current, and output voltage falls.

OIL IN WATER SENSOR

A useful application of photo-cells is in detection of oil-water interface (as an alternative to the method described in Chapters 3 and 13). Fluid passing through glass is exposed to long wavelength light from an ultra violet lamp which causes fluorescence if oil particles are present. This light can be detected by the secondary element photo-cell unit which produces a signal for amplification. The amount of fluorescent light is dependent on the amount of oil in the oil-water mixture and this affects the amount of visible light detected by the photo-cell. An ultrasonic beam between piezo-electrical crystals across the interface is also used, sometimes utilising beam reflection or refraction across the interface.

SMOKE DENSITY DETECTOR

For fire warning and exhaust gas indication a photo-cell in conjunction with an amplifier and alarm or indicator is used. Three types are in use, those which operate by light scatter, by light obscuration and a combination.

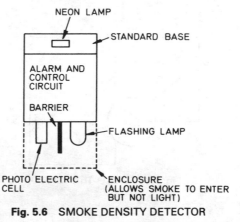

Fig. 5.6 SMOKE DENSITY DETECTOR

A light scatter photo-cell separated from a semi-conductor intermittently flashing light source is shown in Fig. 5.6. The housing enclosure allows smoke but not light inside. With smoke present in the container light is scattered around the barrier on to the photo-cell and an alarm is triggered. The light obscuration type is used in oil mist detection for IC engine crankcases and the combination type is used for detection in CO_2 flooding systems.

OIL MIST DETECTOR

The photo-cells of Fig. 5.7 are normally in a state of electric balance, *i.e.* measure and reference tube mist content in equilibrium. Out of balance current due to rise of crankcase mist density can be arranged to indicate on a galvanometer which can be connected to continuous chart recording and auto visual or audible alarms. The suction fan draws a large volume of slow moving oil-air vapour mixture in turn from various crankcase sel-

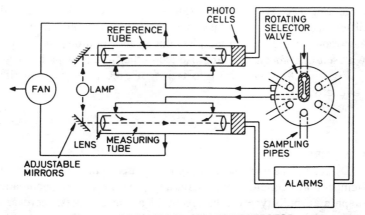

Fig. 5.7 CRANKCASE OIL MIST DETECTOR

ection points. Oil mist near the lower critical density region has a very high optical density. Alarm is normally arranged to operate at 2·5% of the lower critical point, *i.e.* assuming 50 mg/l as lower explosive limit then warning at 1·25 mg/l.

pH SENSOR

The pH value of a solution is the logarithm of the reciprocal of the hydrogen ion concentration in the solution. Its value ranges between 0 to 14, neutrality being 7 anything from 7 to 14 is alkaline and from 0 to 7 is acidic. pH measurement and control is extremely important, being primarily used for feed water analysis.

The method of pH measurement is by means of a conductivity cell consisting of two electrodes and a temperature sensor, pH value varies with temperature hence it is important that this be controlled by means of a sensor/compensator. One of the electrodes is a reference electrode which has a fixed potential irrespective of the variation of hydrogen ion concentration of the solution. The other electrode produces a potential dependent mainly upon the difference in hydrogen ion concentrations between the buffer solution and the solution whose pH has to be measured (across the membrane). In this way the potential difference between the glass measuring electrode and reference electrode is a measure of the pH value of the solution. Electrodes, with sensor/compensator, are inserted in the fluid flow path.

Figure 5.8 shows the two types of electrode used in the conductivity cell.

HEAT (FIRE) DETECTOR

Detector heads are generally one of three types. At a set heat (temperature) condition the increasing pressure on the pneumatic diaphragm bulb type closes electrical alarm contacts, increased differential temperature on the bi-metallic type activates alarms, increased heat fractures a quartzoid bulb (containing a highly expansive fluid) releasing water sprinkler supply and pressure alarm. A typical fire detection-alarm circuit is shown in Fig. 5.9.

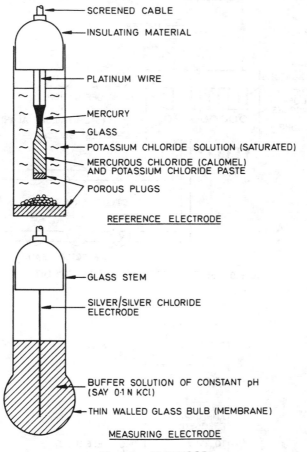

REFERENCE ELECTRODE

MEASURING ELECTRODE

Fig. 5.8 pH SENSOR

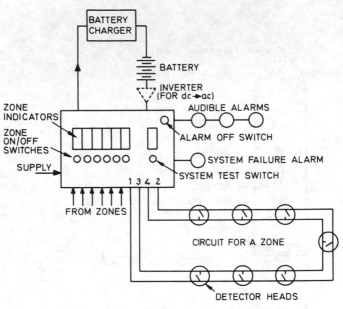

Fig. 5.9.   FIRE DETECTION – ALARM CIRCUIT

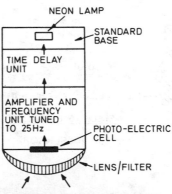

Fig. 5.10   FLAME DETECTOR

FLAME DETECTOR

Fig. 5.10 illustrates the infra red type of device. Flame has a characteristic flicker frequency of about 25 Hz and use is made of this fact to trigger an alarm. Flickering radiation from flames reaches the detector lens/filter unit, which only allows infra-red rays to pass and be focused upon the cell. The signal from the cell goes into the selective amplifier, which is tuned to 25 Hz, then into a time delay unit (to minimise incidence of false alarms, fire has to be present for a pre-determined period), trigger and alarm circuits.

GAS EXPLOSION-DETECTOR METER

The instrument illustrated in Fig. 5.11 is first charged with fresh air from the atmosphere using the rubber aspirator bulb (A). On-off switch (S_2) is closed together with check switch (S_1) and the compensatory filament (C) and detector filament (D) allowed to reach steady state working temperature. The zero adjustment rheostat (F), can now be adjusted so that galvanometer (G) reads zero. Voltage is adjustable from battery (B) by the rheostat (E). Switch S_2 is now opened.

The instrument is now charged from the suspect gas space and while operating the bulb, the switch S_2 is again closed. If a flammable or explosive gas is present it will cause the detector filament to increase in temperature. This disturbs the bridge balance and a current flows. Galvanometer G can be calibrated so that the scale is marked to read "% of Lower Limit of Explosive Concentration of Gas".

An alternative design has two ionising chambers, one reference (air) and the other sample, each containing a radioactive ionising source. Combustion particles when ionised are more bulky and less mobile than normal gas molecules so they are readily neutralised. This results in higher resistance and voltage change at the sample chamber — which activates alarms.

GAS ANALYSIS

For detailed gas analysis an Orsat apparatus is used. However a number of measurements require to be continuously recorded. Two representative examples can be considered namely an oxygen analyser and a carbon dioxide analyser.

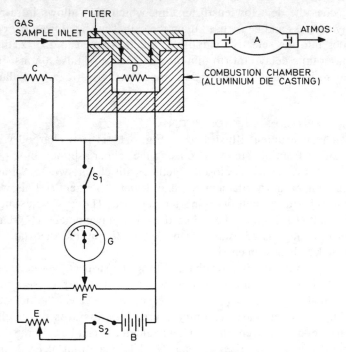

Fig. 5.11   GAS EXPLOSION – DETECTOR METER

Oxygen analyser

Gases can generally be classified as either diamagnetic or paramagnetic, the former seek the weakest part of a magnetic field and the latter the strongest. Most of the common gases are diamagnetic but oxygen is paramagnetic and use is made of this in the oxygen analyser shown in Fig. 5.12.

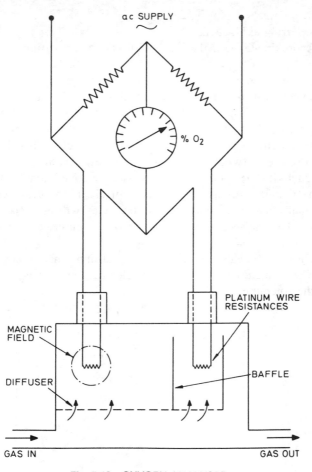

Fig. 5.12 OXYGEN ANALYSER

Two platinum wire resistances are heated by current from an ac bridge and the gas to be measured enters the resistance chamber via a diffuser. One of the resistance wires is placed in a magnetic field hence oxygen is drawn towards this resistance, thus convection currents are set up around this resistance which is then cooled relatively to the other resistance. The bridge is then unbalanced, the amount of unbalance is a measure of the oxygen content and this is displayed on the galvanometer.

CO_2 *analyser*

Referring to Fig. 5.13.

Approximate thermal conductivities are in proportion:

$$CO_2 = 1, H_2O = 1, CO = 4, O_2 = 2, N_2 = 2$$

The sample enters via a filter and drier, water vapour must be removed as it has the same conductivity as CO_2. The wire cell resistance is proportional to heat dissipation, proportional to thermal conductivity of gas in the cell, proportional therefore to CO_2 content. Air is used in the reference cell. Thus the only difference between gas sample and air, from the thermal conductivity viewpoint is CO_2 (as H_2O removed and O_2 and N_2 same value). This assumes no CO or H_2, if these are present (normally only very small proportions) they will be registered as CO_2 unless the sample is first passed over a burner and these two gases burned off before the reading.

Thus the Wheatstone bridge electrical unbalance is dependent on CO_2 content and the unbalance electrical current is measured by the potentiometer.

Chemical absorption and mechanical types are also used.

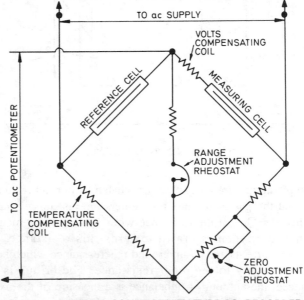

Fig. 5.13 THERMAL CONDUCTIVITY TYPE CO_2 RECORDER

RELATIVE HUMIDITY

A hair element will react to changes of humidity and provide a linear movement, with negligible force, which can be converted to electrical or pneumatic signal and amplified as required.

WATER ANALYSIS

The measurement of pH has been considered previously. Two other measurements are commonly required, *i.e.* electrical conductivity meter for dissolved solid assessment and dissolved oxygen meter.

"Dionic" water purity meter

Specific conductivity mho/cm^3 is the conductance of a column of mercury 1 cm^2 cross sectional area and 1 cm long. This is a large unit and micromho/cm^3 (reciprocal megohm) is used and when corrected to 20°C is called a dionic unit. Conductivity of pure distilled water is about 0·5, fresh water about 500 dionic units. The sensor, shown in Fig. 5.14, measures conductivity of

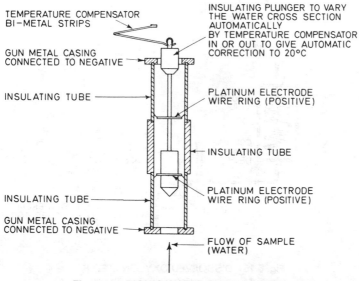

Fig. 5.14 DIONIC WATER PURITY METER

two water columns in parallel, *i.e.* between positive platinum rings and negative gunmetal collars. The insulating plunger, operated by bi-metallic strip, varies cross sectional area for automatic correction to 20°C. The measurement is by conventional ohmeter. The device should be used with de-gassifying units to avoid errors due to occlusion of carbon dioxide.

Dissolved oxygen meter

The unit is shown in Fig. 5.15. The sample water flows via a chamber which surrounds the katharometer (Wheatstone bridge circuit) and receives pure hydrogen. Some hydrogen is taken into solution and this releases some dissolved oxygen (in air). This mixture passes to atmosphere across one side of the bridge whilst the other side is in pure hydrogen. The cooling effect is different on the two sides of the katharometer, depending on air (oxygen) present, and resultant unbalance current operates indicator or recorder calibrated directly in ppm oxygen. For very low oxygen content it is often necessary to utilise an electro-chemical cell in place of this meter.

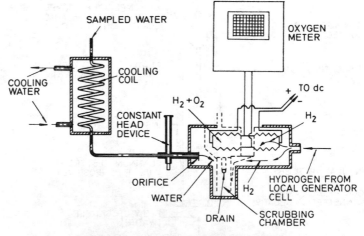

Fig. 5.15   DISSOLVED OXYGEN METER

INSTRUMENT CALIBRATION. TESTING AND ADJUSTMENT

Generally a specialist subject. Pneumatic instruments would be tested by master gauge, standard manometer or deadweight tester. Electrical instruments are tested by standard resistors, potentiometers, capacitors, etc.

Using a Bourdon pressure gauge as example:

1. Zero (error) adjustment changes base point without changing the slope or shape of the calibration curve. It is usually achieved by rotating the indicator pointer relative to the movement, linkage and element

2. Multiplication (magnification) adjustment alters the slope without changing base point or shape. This is effected by altering the drive linkage length ratios between primary element and indicator pointer.

3. Angularity adjustment changes the curve shape without altering base point and alters scale calibration at the ends. This error is minimised by ensuring that link arms are perpendicular with the pointer at mid scale.

Fig. 5.16 shows calibration curves and adjustment for the Bourdon link type of instrument mechanism.

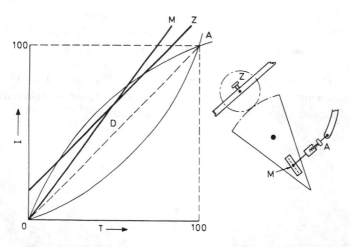

Fig. 5.16 INSTRUMENT CALIBRATION AND ADJUSTMENT

Instrument readings (I), true values (T), desired result (D). Zero error and adjustment (Z), multiplication error and adjustment (M), angularity error and adjustment (A) — error curves or lines for actual values.

Over the design range pointer movement bears a linear relationship to pressure, and the scale is calibrated accordingly.

Hysteresis — a vibration phenomena. Best eliminated by correctly meshed gearing and fitted pivots to reduce backlash, etc.

TEST EXAMPLES 5

1. Explain the principle of operation of a carbon dioxide recorder for monitoring the uptake gases.
State what normal maintenance it requires.
State what action is taken if the carbon dioxide content is unacceptably low.
If this action does not alter the carbon dioxide reading, explain how the accuracy of the recorder is checked and adjusted.

2.   Explain the term "photo-electric effect" and describe equipment suitable for crankcase monitoring and fire detection in which this phenomenon is utilised.

3. Describe, with the aid of sketches, a torsionmeter.
Explain the principle of operation and briefly describe the construction of the shaft and indicator units.

4.   Discuss the errors liable to be exhibited by link type instruments. Describe how such an instrument could be calibrated and adjusted to reduce these errors to a minimum.

CHAPTER 6

TELEMETERING

The objective of this chapter (and Chapter 7) is to link instrumentation in preceding chapters, primarily concerned with measuring devices, with control elements as described in subsequent chapters. Some repetition may inevitably result in the presentation.

There are a wide range of variables to be measured. Detecting and measuring elements are mainly electrical but a significant number are displacement operated mechanical types. Chemical and electronic devices are also used.

Transducers can generally be simplified into three basic reversible types namely:

$$\text{mechanical displacement} \leftrightarrow \text{pneumatic}$$

$$\text{mechanical displacement} \leftrightarrow \text{electrical}$$

$$\text{pneumatic} \leftrightarrow \text{electrical}$$

Pneumatic principles are invariably flapper nozzle; orifice, diaphragm. Electrical principles include resistance change, variable inductance, variable capacitance, current or voltage, with frequency and phase used to a limited extent. Conversion of electrical signal is also used, *i.e.* resistance-current, voltage-current, etc. and such modern transducers often incorporate electronic oscillators and amplifiers.

Telemetering may be defined as signal transmission over a considerable distance. The device at the measure point, usually a transducer, is then often called a transmitter with the receiver located at the recording or control centre.

The material in this chapter will be covered in three sections namely pneumatic transmitters, electrical transmitters, receivers.

PNEUMATIC TRANSMITTERS
POSITION-BALANCE TRANSDUCER (PNEUMATIC)

Displacement of a mechanical linkage gives variation in pneumatic signal output pressure. The flapper-nozzle is the basis of many pneumatic mechanisms and the position (motion) balance is essentially a *balance* of positions (see Fig. 6.1).

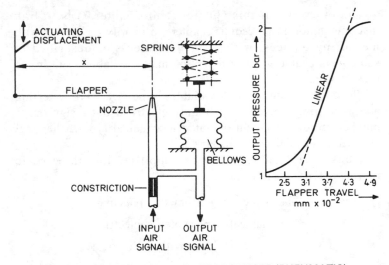

Fig. 6.1 POSITION BALANCE TRANSDUCER (PNEUMATIC)

Ideally equal increments of flapper movement should produce equal increments of pressure output, *i.e.* linear proportionality. In practice this is only achieved over a limited flapper travel. To ensure increased sensitivity and linearity negative feedback is used via a bellows. Linear output over the pressure range is obtained for an effective flapper travel range near the nozzle of about 0·015 mm. Output signal pressure is proportional to actuating link travel and the device is adjustable by varying the mechanical advantage of the flapper lever, *i.e.* altering the *x* dimension to the right or the left. A typical output pressure-flapper travel characteristic is included in Fig. 6.1. The device is obviously a displacement-air pressure transducer, displacement variation from flow, level, etc. variables. A pneumatic relay can be fitted on the air input.

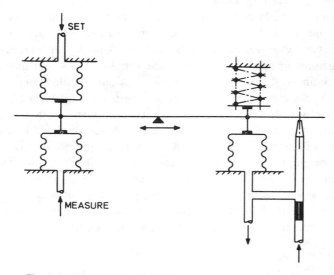

Fig. 6.2 FORCE BALANCE TRANSDUCER (PNEUMATIC)

FORCE-BALANCE TRANSDUCER (PNEUMATIC)

This is essentially a *null* method, *i.e.* equal and opposite forces (torques) which eliminates inherent errors of the position-balance device. Consider Fig. 6.2, the flapper is a constrained bar pivoted about a fulcrum (adjustment of which varies mechanical advantage and thus changes proportion of input to output change). Bellows have equal effective area. With the device in equilibrium assume an increase in measured signal pressure which will produce a net up force and clockwise torque on the bar. The flapper movement towards the nozzle will continue until the increased output pressure in the feedback bellows produces an anti-clockwise torque, to balance the actuating torque, and at this point equilibrium is restored. Again a displacement variable to air pressure transducer. A relay may be fitted.

ELECTRO-PNEUMATIC TRANSDUCER

The unit shown in Fig. 6.3 is based on the force balance principle with input variable of current (10-50 mA dc is usual). Electrical current signal variation causes a torque motor to produce a variable force (F) which is balanced by the feedback pneumatic bellows force (B) at equilibrium. The bar is circular and between the poles of the permanent magnet acts as an armature when excited by a dc current. Consider an increase in armature current; the strength of the armature poles will increase accordingly. The S pole will move up as unlike poles attract and produce a clockwise moment about the pivot whilst the N pole will move down producing an anti-clockwise moment. The moment arm of the S pole force is greater so there is a net clockwise moment. This action closes in on the nozzle giving a higher output pressure and increasing the feedback bellows force until equilibrium is achieved, *i.e.* a direct acting transducer. An alternative torque motor design utilises a suspended coil within the magnetic field of a permanent magnet. Fig. 6.3 is an input electrical (current) to output pneumatic pressure transducer (I/PP).

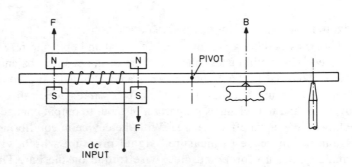

Fig. 6.3 ELECTRO-PNEUMATIC TRANSDUCER

ELECTRICAL TRANSMITTERS

Consider first three examples of position (motion) balance converters namely variable resistance, inductance and capacitance.

VARIABLE RESISTANCE TRANSDUCER

In this simple and early form of transducer the mechanical movement of the measuring element connection varies the electrical circuit resistance. The device is restricted in use due to contact problems, variations in wire connection resistance and the need for a constant stabilised voltage supply. Two types are shown in Fig. 6.4, *i.e.* mV measurement and mA measurement locally or remote, the latter is a potentiometer technique.

Displacement may be due to any variable variation, *i.e.* level, pressure, etc. via Bourdon tube, diaphragm or similar displacement device. Provision of a cross coil measuring meter, or remote receiver, makes indication virtually independent of supply voltage variations. One coil is directly energised from supply (current proportional to voltage) and a cross coil is deflected by measurement current so that deflection of the meter within the permanent magnet is proportional only to resistance of the transmitter. This is a two-coil ratiometer.

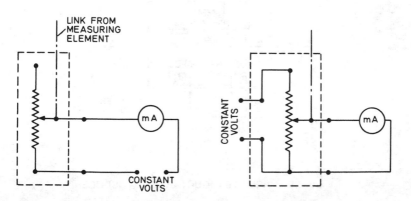

Fig. 6.4 VARIABLE CONTACT RESISTANCE TRANSDUCER

OTHER RESISTANCE SYSTEMS

The Wheatstone bridge principle is used in many cases, both dc and ac, with direct or null balance techniques. The bridge together with electrical resistance sensors for resistance thermometers (Chapter 1) and strain gauges (Chapter 2) have been described in detail already. Another electrical resistance transmitter has also been discussed previously in Chapter 3 namely the type utilising electrical resistors in mercury for measurement of level.

VARIABLE INDUCTANCE TRANSDUCER

The unit shown in Fig. 6.5 is a differential transformer with three coils fully wound on a cylindrical former (only half the winding is shown on the sketch for simplicity). The core, which is moved laterally by displacement of a sensor element, provides the magnetic linking flux path between coils. The primary ac voltage induces secondary voltages and as the two secondary windings are in series opposition the two outputs are opposite in magnitude and phase with the core laterally in the middle of the former.

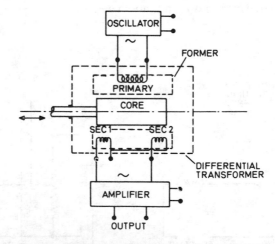

Fig. 6.5   VARIABLE INDUCTANCE TRANSDUCER

With the core moved right the induced voltage in secondary 2 winding increases and that in secondary 1 decreases so giving a differential output. Similarly movement left gives a voltage difference but 180° out of phase. The characteristic can be linear with zero volts at mid travel. Input displacement, from such as Bourdon tube or diaphragm is converted to an electrical signal for telemetering to indicators, recorders, data processing or electronic controllers.

Many electrical transducers are combined in the transmitting unit with oscillators and amplifiers of solid state modular assemblies. The oscillator supply in Fig. 6.5 from a power supply unit is commonly at 12V dc and incorporates a chopper unit, ac amplifier and output-feedback stage. A stabilised current to the primary winding of about 5 mA is often used from say a 1·6 kHz oscillator. The amplifier itself, 12V dc supply, accepts ac input from the differential transformer via ac bridge circuits at up to 2 mA and gives output via demodulator-filter circuits at about 50 mA maximum, dc. It is effectively a dc input dc output system. Components mentioned above are considered in more detail in the next chapter.

This *inductance ratio* system can employ any type of receiver. For a simple system direct ac supply, without oscillator and amplifier, can be used and ac output passed across a bridge rectifier for each coil to a two-coil ratiometer with pointer indicator.

In an *inductance balance* system the receiver is identical to the transmitter with secondary windings interconnected. Unbalanced emfs due to displacement and inductance change at the transmitter result in corresponding displacement at the receiver so as to maintain current in each part of the circuit constant.

(The lever and gearbox are mechanical transformers.)

VARIABLE CAPACITANCE TRANSDUCER

The capacitance of a parallel plate capacitor is given by

$$C = \frac{A\varepsilon}{d}$$

A is plate area, ε absolute permittivity and *d* plate separation. Change of capacitance utilisation in conjunction with an ac bridge circuit has been considered previously with reference to level probes (Chapter 3). For displacement measurement a parallel RL resonant circuit can be utilised. Alternatively a differential capacitor principle can be used for displacement-current conversion, and is now described.

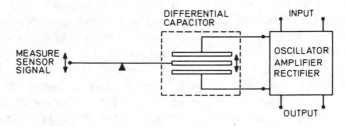

Fig. 6.6   DIFFERENTIAL CAPACITOR TRANSDUCER

Consider Fig. 6.6:

The central plate of the differential capacitor is moved vertically by the displacement of a sensor device. The outer plates are connected to a combined oscillator, amplifier, rectifier unit.

Movement of the centre plate, towards one fixed plate and away from the other gives a change in capacitance to the oscillator-amplifier. A change in output current to receiver results.

ELECTRONIC FORCE-BALANCE SYSTEM

An electro-pneumatic converter has already been described and an alternative pneumatic (pressure, force, displacement, etc.) electric converter can now be considered (Fig. 6.7).

When movement varies the inductance, coupled to the oscillator amplifier, let us assume amplifier output current increases. This will continue until the feedback current on the force motor produces equilibrium. Effective full scale beam travel is only about 25 microns. Input may be from Bourdon tube or diaphragm. The device is referred to as a PP/I transducer.

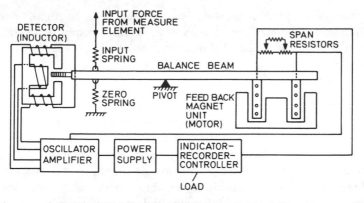

Fig. 6.7 ELECTRONIC FORCE BALANCE SYSTEM

VOLTAGE-CURRENT TRANSDUCER

It is often necessary to use a mV/I converter when dealing with thermocouple or resistance thermometer inputs. Such a device is shown in Fig. 6.8.

Deviation between input mV and a standardised zero suppression voltage, from a zener diode power pack and bridge, is algebraically added and passed through a filter network (to avoid stray ac pick-up). This signal is algebraically added to the feedback stabilising loop of the amplifier and passed through the amplifier to output. Adjustable resistors A, C, D allow temperature correction, zero adjustment and span control of feedback; B

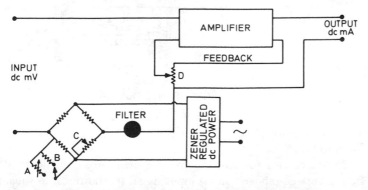

Fig. 6.8 VOLTAGE-CURRENT TRANSDUCER

measured value. A dc amplifier can be used. If an ac amplifier is preferred a chopper input and synchro-rectifier output is needed, output and feedback isolated by a transformer.

RECEIVERS

The variation of receiver types is very large ranging from direct measuring meters, recorders, display units, controllers and analysing units. In many cases if a transducer is used it is merely a form of converter device as already described in this chapter. Frequently it is not possible to sensibly separate transmitter and receiver because they are inherently linked in operating principle. With these provisos in mind a selection of units not previously, or subsequently, described are now presented.

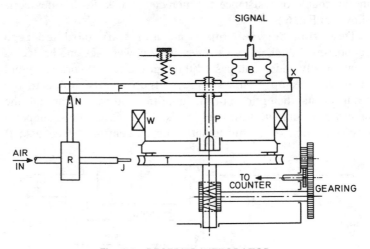

Fig. 6.9 RECEIVER INTEGRATOR

RECEIVER INTEGRATOR
Refer to Fig. 6.9:

This receiver is for flow recording. Down force on bellows (B) from increased (above datum) input signal (proportional to square of flow measurement) and about fulcrum (X) closes force bar (F) into (F) into nozzle (N). Increased pressure from relay (R) acts on a 60 "tooth" turbine wheel (T) to cause rotation and equilibrium is obtained when up force due to centrifugal force (proportional to square of turbine wheel speed) through thrust pin (P) on bar balances down force. Adjustment is via spring (S) causing movement of weights (W) (up increases feedback force and reduces counts per unit input pressure).

Turbine wheel speed is directly proportional to flow, and by gear reduction to the counter, indication is of total flow

POTENTIOMETRIC PEN RECORDER

The potentiometer is used a great deal in instrument systems and also control systems (position control, Chapter 14). For recording of small dc voltage it is usual to convert to a suitable frequency ac and amplify although dc amplification can be used. Conversion to ac is achieved in a dc chopper amplifier with output direct to servo-amplifier pen drive motor.

Figure 6.10 is a simplified sketch of a continuous balance system.

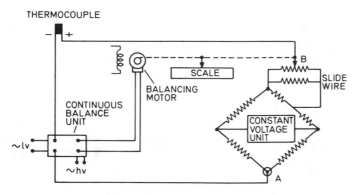

Fig. 6.10 POTENTIOMETER CIRCUIT

Input dc voltage, from say a thermocouple, is measured against slidewire voltage at B with a constant voltage bridge source. Difference between A and B is amplified at the continuous balance unit so energising the balancing motor to move pen arm and B until the voltage difference is zero. Similar arrangements utilise conductivity, ratio, bridge, etc. circuits. The balancing motor is two phase with a reference winding and a control winding from the balance unit. Input to the balance unit incorporates a converter and centre tap of an input transformer. The vibrating reed converter, in moving between two contacts, allows current to pass alternately through each half of the transformer. Secondary ac voltage is amplified and fed to the control winding of the balancing motor, so timed with ac supply to give the correct restoring action. The chart is driven by a constant speed geared motor. A damping feedback tachogenerator driven from the balance motor is often fitted.

XY RECORDER

Used to measure a quantity *Y*, varying with *X*, where *X* is not a function of time. Two servo systems, perpendicularly connected, cause the pen to move to any area position on the chart. Inputs cause perpendicular travel related to *X* and *Y*.

POSITION MOTORS (dc)

One technique is to feed current into a toroid resistor transmitter with three tappings connecting to a three phase star winding enclosing a two pole rotor receiver. This is sometimes referred to as a Desynn transmission link. A similar principle is used in the position indicator (electric telegraph).

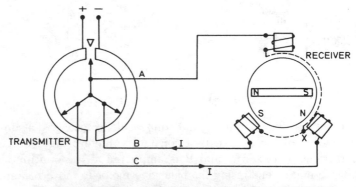

Fig. 6.11   POSITION INDICATOR (ELECTRIC TELEGRAPH)

Figure 6.11 shows the arrangement in equilibrium with equal currents (I) in line B and C and zero current in line A. The receiver rotor is locked by equal and opposite torques from the attractions on unlike pole faces. Assume the transmitter to be moved 30° clockwise. Current flows to receiver from line C, subdivides at point X and equal currents return through lines A and B of magnitude I/2. This creates a strong N pole at fixed magnet X and two weak S poles at the other two fixed magnets. The receiver indicator will therefore turn to the corresponding equilibrium position, *i.e.* 30° clockwise.

POSITION MOTORS (ac)

These devices are usually known by trade names such as synchro, resolver, magslip and for larger powers, selsyn. Figure 6.12 shows a transmitter and receiver of a synchro system.

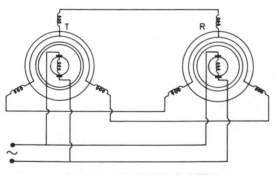

Fig. 6.12 SYNCHRO SYSTEM

Both rotors are supplied from the same ac source and stators are linked in star. With rotors in the same angular position emfs from transmitter and receiver stators balance and there is no circulatory current. If the transmitter rotor is moved, induced emfs are unequal and current circulates so producing a torque to bring the receiver rotor into line and restore equilibrium. Zero receiver torque exists at alignment and maximum occurs at 90° out of alignment. The resolver system is similar but utilises two phase and is used for both fine control and data processing systems. An intermediate synchro (follow through, hunter differential, etc.) can be arranged, with three phase rotor and stator connections, so that summing or differential control outputs are possible.

TEST EXAMPLES 6

1.   In a pneumatic telemetering system explain how the air flow through a nozzle in a transmitter is varied. How does this variation cause a change in the signal to the receiver?

2.   Sketch and describe a variable inductance and a variable capacitance type of transducer. Discuss typical applications, with examples, of these devices in measuring or control systems.

3.   Describe a remote telemetering system. Detail on a suitable diagram both the indicator and receiver unit and the connections between.

4.   Describe, with a suitable sketch, an electro-pneumatic converter. Explain the operation when subject to a change in input. State two applications of such a device.

CHAPTER 7

ELECTRONIC DEVICES

Electronics has developed greatly in this century and especially in recent years. The subject is specialised and a full presentation is available from a wide range of books. For the purpose of this short chapter an extreme rationalisation is necessary as follows:

(1) Components described are only those active devices with a direct application to instrumentation and control and the basic introductory electronics theory is reduced to a minimum.

(2) Description is related to semi-conductor devices only. All the theory and practice of vacuum valve electronics including diode, triode, tetrode, pentode, etc. is generally omitted as obsolete.

The contents of this chapter are given in 5 sections namely; semi-conductors, rectifiers, amplifiers, oscillators and other devices. It applies generally to linear integrated (analogue) circuits in the frequency domain, *i.e.* concerned with sinusoidal (or similar) waveform and frequency analysis (current and voltage). Chapter 16 includes integrated digital circuits in the time domain, *i.e.* concerned usually with square waveforms where the criteria is whether current or voltage is present or not *i.e.* logic devices, computers, etc.

SEMI-CONDUCTORS

ATOMIC THEORY

The basic theory of electronics relates to atomic physics with negatively charged electrons orbiting around the positively charged nucleus. Bonding decides conductivity, for example copper as a good conductor and mica as an insulator. Conventional current flow is positive to negative. When considering electron drift movement, which takes place from the negative terminal through the conductor to the positive terminal, it must be noted that the flow is opposite to the conventional.

Germanium and silicon in pure form have a diamond lattice formation, each of four valence (binding) electrons per atom, and electron flow is only possible by partial lattice breakdown due to thermal energy. Arsenic and antimony have a five valency shell readily fitting into the diamond lattice and leaving a surplus conduction electron. Boron or gallium have a three valency shell and whilst bonding into the lattice occurs a missing electron creates a conduction hole.

ELECTRON CONDUCTION

With *conductors* the electrons, which constitute the current flow, are capable of drifting through the material. The electrons can be imagined to lie in a valance band and energy supply, heat for example, is sufficient to excite the electrons sufficiently to allow them to jump across a narrow non-conducting band into the conducting drift band.

With *insulators* a non-conducting band is wider and electron jump does not readily occur, so that no current flows.

With intrinsic *semi-conductors*, such as germanium or silicon, the properties are midway between conductors and insulators. Some electrons can break from the crystal lattice bond structure, the gap created can then be filled by another electron, hence electron movement does occur and this increases with temperature increase. Extrinsic *semi-conductor* materials contain slightly impure or doped material. if arsenic (antimony, phosphorus) is added to germanium or silicon there is a surplus of electrons in the crystal lattice, such *donor* atoms give an *n*-type conductor. If boron (gallium, indium) is added there is a positive *gap* or *hole* in the crystal lattice, such *acceptor* atoms give a *p*-type conductor.

Conduction by electrons is similar to that in metals. For a *p*-type material, an electron moves through the lattice, being attracted by the positive hole, to fill the hole, this creates another hole, and so on. The hole then appears to move to the negative through the lattice, this can be regarded as causing the conduction. Majority carriers, electrons or holes in n-type and p-type impurities, greatly outnumber minority carriers (which are temperature excited electrons). The temperature (maximum) for solid state devices is usually fixed about 75°C (above which thermal runaway occurs). Aluminium is used for contacts.

When a p-type and an n-type material are made into a specific junction electron flow to the holes in the p-type occurs. A negative charge exists in the p-type and a positive charge in the n-type, thus giving a potential difference across the junction which stops further migration of electrons. This **n-p junction** can act for example as a rectifier, with the positive side connected to the n-type the potential difference is increased, with the negative side connected to the n-type the potential difference is reduced. A few random electrons (leakage current) can go against the bias. When reverse biased the holes and electrons are drawn away from the junction leaving a depletion layer with virtually no current carriers.

Further consideration leads to transistors in which a *pnp* or *npn* sandwich exists. For *npn* the emitter emits electrons, collector collects electrons, and the base controls the flow of electrons by

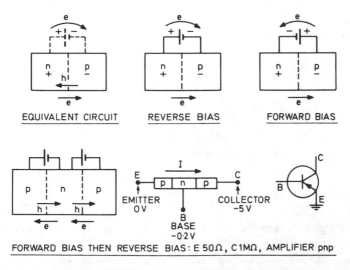

Fig. 7.1 SOLID STATE JUNCTIONS, EXAMPLES

controlling the charge concentration in the base region. For *pnp* the polarity is reversed and the flow is as shown on Fig. 7.1. A relatively larger emitter-collector current can be controlled by a small base-collector current and voltage.

Current gain (collector to base current ratio) can easily be in the range 10 to 200. Power gain may be as high as 50 000.

Silicon planar technique (diffusion) is the basic process used in bipolar (two carriers e and h) solid state and integrated circuit chip technology for resistors, capacitors, diodes and transistors in circuit.

RECTIFIERS

Metal rectifiers, dependent on barrier layer properties, such as copper-oxide and selenium are in general use. In the heavy electrical industry mercury arc and ignitron units are employed. For electronics the valve diode is well known as a rectifier and for larger power, in conjunction with control applications such as energising (firing) relays etc., the gas-filled grid design is used – thyratron. The latter two have generally been superseded by the semi-conductor diode (*pn* junction rectifier) and the thyristor (silicon controlled rectifier). More complex devices such as tunnel, variable capacitance, microwave and four layer diodes are not considered in this book.

The bridge rectifier is shown in Fig. 7.2.

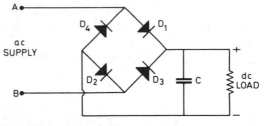

Fig. 7.2   BRIDGE RECTIFIER

When A is positive with respect to B current flows through D_1 to the load and returns through D_2 to B. When B is positive with respect to A flow is through D_3 and returns through D_4.

The centre tap transformer rectifier is shown in Fig. 7.3

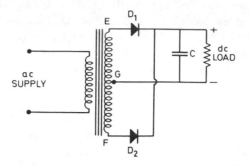

Fig. 7.3. CENTRE TAP TRANSFORMER RECTIFIER

When E is positive with respect to F current flows through D_1 to the load and returns to G, D_2 is reverse biased. When F is positive with respect to E flow is via D_2 to the load, returning to G, D_1 reverse biased

APPLICATIONS

Low frequency (audio) signals cannot be efficiently radiated. A high frequency (radio) **carrier** has a low frequency signal impressed on it at the transmitter by varying or **modulating** the amplitude of the carrier wave in sympathy with the low frequency signal. At the receiver the signal information is recovered from the carrier wave by a process known as **demodulation** or detection. The wave at the receiver is first rectified (detected) and the signal information recovered by passive networks (resistance-capacitance-inductance). These networks are usually known as **smoothing** or **filter** circuits. Figure 7.4 shows a transmitter-receiver

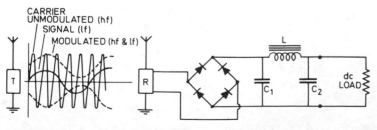

Fig. 7.4 DEMODULATION CIRCUIT

unit with full wave bridge rectification-detection by semi-conductor diodes and a high frequency capacitor (C_1) filter with smoothing choke inductor (L) and filter capacitor (C_2). This is a capacitor input filter system, C_1 is often termed reservoir and C_2 smoothing capacitor. A capacitor provides high impedance to dc and an inductor high impedance to ac. The function of the filter may also be explained by regarding it as an integrating network. In general conversion between ac and dc is often required in instrument-control systems. Modulation, demodulation and re-modulation are frequently used in the electronic systems to utilise the best component for a particular duty of sensing, amplifying, control, etc. Low voltage supplies are often required for transistorised equipment. Typically a mains transformer, bridge rectifier, smoothing circuit and diode (zener) volts stabilisation is used.

SEMI-CONDUCTOR DIODE

When ac current is applied to the *pn* junction a large current will flow when forward biased polarity applies and conduction stops when reverse biased and the diode acts like any other rectifier. This is confirmed by the characteristic shown in Fig. 7.5

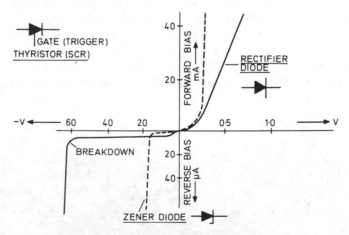

Fig. 7.5 pn JUNCTION CHARACTERISTIC

If the reverse bias is increased beyond a certain value, called the breakdown voltage, the reverse current will increase sharply and this is known as avalanche or zener current. Rectification diodes never operate in the breakdown region but the zener diode, used in voltage stabilisation and reference circuits, is operated in this region. Note in particular the symbols shown in Fig. 7.5 to represent semi-conductor rectification, zener diodes, thyristor (silicon controlled rectifier) and the *different* scales on the axes. Current flow is in the direction of the arrow, the bar to indicate non-reversal (see also Fig. 7.2, bridge rectifier circuit).

ZENER DIODE

Figure 7.6 shows two applications of zener diodes. As a voltage regulator (stabiliser) the reverse connected zener diode conducts if input voltage is above breakdown voltage and current from the supply is the sum of diode and load current. For input voltage increase then current increases through both R and the diode but the diode resistance decreases and current through the diode further increases. A larger volt drop across R will occur but output voltage across the diode remains reasonably constant. Variations of input or output cause shunt of more or less current through the diode resulting in constant voltage, *i.e.* across the diode regulation circuit. (see also Fig. 11.9).

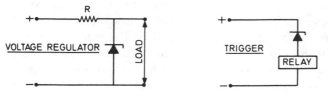

Fig. 7.6   ZENER DIODE APPLICATIONS

The second diagram illustrates use as a trigger safety device. The relay will be held and will not fire until a certain prescribed voltage is reached.

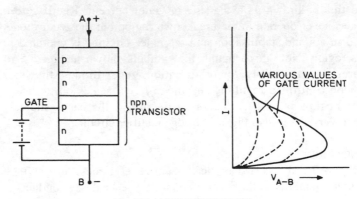

Fig. 7.7 THYRISTOR (SCR)

THYRISTOR (SILICON CONTROLLED RECTIFIER)

From Fig. 7.7 it will be noted that this is a *pnpn* sandwich. Consider the section shown as an *npn* transistor which requires a base current from the gate to cause conduction from collector to emitter. With A negative both top and bottom junctions are reverse biased. With A positive and no gate voltage the middle junction is reverse biased. As shown in Fig. 7.7, and with a pulse of positive current injected into the gate, then as the *pn* junction (top) is already forward biased the device is turned on. When conduction exceeds about 10 mA an avalanche (zener) effect occurs and gate current is not required to maintain flow. Operation is similar to the grid thyratron except that it is fired by current and not voltage. To stop conduction, voltage has to be reduced to zero and the device requires another signal to fire. Applications include inversion, stabilisation, regulation. High currents can be controlled and the device is a very fast, high power gain, switch. Small gate current can switch a large load current or gate controlled half wave rectifier. A triac has two SCRs back to back with one gate controlling conduction in either direction. A diac (no gate) builds up voltage to trigger pulse the triac gate current. An application of the thyristor is given in Chapter 14.

AMPLIFIERS

Amplifiers are an essential part of instrumentation and control. The pneumatic amplifier (relay) is described in Chapter 10. Thermionic valve amplifiers (triodes) have been extensively used although the solid state junction transistor is now preferred. Feedback is an inherent addition but a detailed discussion on this topic is reserved until the end of this chapter. Devices are active, *i.e.* utilising an external power source – as distinct from passive.

ROTATING ELECTRICAL AMPLIFIERS

The separately excited generator is often used in control systems. The Ward Leonard type unit is generally well known and an application and description are given in Chapter 14. For higher sensitivity the cross-field dc generator, such as amplidyne and metadyne, can be used. A pair of brushes at 90° to the main brushes provides output and the main brushes are short circuited. A small input signal current to the control winding gives a large change of current in the short circuit and hence the load circuit.

MAGNETIC AMPLIFIER

An amplifier of interest because of its use in electronic components, often in series with transistor amplifiers. A high permeability ferromagnetic core is wound with an ac current coil.

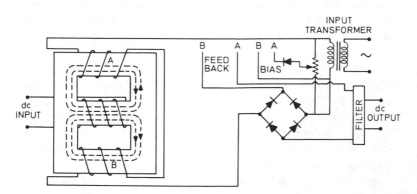

Fig. 7.8   MAGNETIC AMPLIFIER

The inductance opposes increases of current until the induced magnetic flux saturates the core when the reactance now behaves as a resistor, *i.e.* saturated reactor and large increases in current can occur. A dc energised control winding on one limb of the core brings about, and varies, the degree of saturation and hence ac current flow in the gate limb. Bias and feedback windings are also incorporated on gates to improve flexibility and stability, and ac output can be rectified to dc which is filtered and appreciably amplified above the input. The typical unit is shown in Fig. 7.8; ac input is from a transistorised oscillator and dc input is stabilised.

Flux due to dc is unidirectional through gate windings A and B but flux due to ac is, as shown, in opposite directions. The ac output can be passed through a bridge rectifier and filter. Positive feedback is utilised for gain adjustment without oscillation problems because this unit is inherently very stable. The principle can also be used in a transducer with sensor core to vary inductance by movement.

CLASSIFICATION

Generally this depends on duty. If frequency is the criterion the range is from zf (zero frequency – dc amplifier) through lf (audio), rf, vhf to uhf (900 MHz). Generally the first two are of interest here (see Fig. 7.20).

Another classification depends on the equipment to be controlled by the amplifier. Voltage types give undistorted voltage output and power amplifiers are to provide drive power. The latter is used to provide power to recorders, controllers, etc.

A third criterion depends on the position of the bias point in relation to the characteristic curve of the device. If operated near the middle of the linear characteristic there is no distortion and this is termed a Class A amplifier, which is ideal for voltage amplification but has a low power efficiency. If biased at or near "cut off" this is termed Class B and has a good power efficiency but severely distorted output – often half wave. Two such amplifiers can be matched as a push-pull amplifier, which via a transformer gives undistorted output. Class C amplifiers are biased past cut off and have the highest efficiency and greatest distortion (reverse biasing).

With transistors the device is usually utilised in one of two modes *linear* and *non-linear*. In the former sinusoidal signals are amplified without distortion usually in two main types, *i.e.* small signal voltage amplification and power amplification. The non-linear mode utilises switching from off to saturated condition very rapidly. Typical applications include: oscillators, *i.e.* square (or saw) wave supply generators, bistable (flip-flop) devices used in counting circuits, etc.; static switching and hold (memory) circuits, logic devices, etc. These applications are described later in this chapter and in Chapter 16.

JUNCTION TRANSISTOR

The *pnp* junction is shown in Fig. 7.1. Due to initial diffusion, one junction (emitter-base) is forward biased and the other junction (base-collector) is reverse biased. The former conducts heavily and positive carriers (holes) diffuse across the n region. If this region is arranged to have few electrons little combination occurs and holes are attracted across the next junction by the bias voltage. Emitter current is the sum of collector current and base current. The collector current is much greater than the base current, and proportional to it over a wide range, so that collector current can be controlled by the base current. Power for emitter-collector current does not come from the amplifier input current, *i.e.* base current, but from an external power source. This amplifier is an active device and in system terms can be treated as a "black box", *i.e.* input, output and power source relationships required without details of internal "box" arrangements. (For *npn*, biasing voltages are reversed and conduction is mainly due to electrons). ***pnp* is used in this text.**

Factors of importance are transfer characteristic (function) (comparison of input and output), dynamic range (of power), efficiency (on power basis), amplification (gain) (magnitude of input and output voltages or currents) and frequency response (transit times).

The transistor acts like a triode value, except that no current flows in the grid to cathode circuit of the valve but current does flow in the base-emitter circuit of the transistor. The input circuit is biased to produce steady base current flow. Changes in input current cause much larger changes in output current, i.e. current amplifier. It is a bipolar (utilises both charge carriers) device.

Circuit Configurations

Means to connect input, supply power, and to make use of output are necessary. There are three ways to connect, differing in the way input and output are connected to the transistor, and each gives different characteristics which will depend on duty. Power must be consumed in the external circuit (load), usually by resistors, and the collector current passes through the transistor and load resistor.

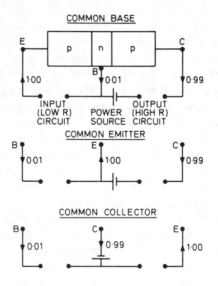

Fig. 7.9   CIRCUIT CONFIGURATIONS

Configurations are as shown in Fig. 7.9. These are common base, common emitter and common collector. This depends on the electrode which is common to the input and output signal; common emitter is the most often used. From Fig. 7.9, for the common base connection, assuming an ac input of say 45mV and 1kΩ resistor in the collector circuit, then 990 mV will be developed across the resistor, i.e voltage gain 20. For the common emitter connection the input circuit has a much lower current, therefore the input resistance is about 100 times higher and without the resistor the current gain is 99. The common collector has similar characteristics to the latter. Currents shown in Fig. 7.9, in mA are illustrative.

Transfer Characteristics

These depend on circuit arrangement adopted as well as upon the transistor itself. Consider the junction transistor as a current amplification device. The current transfer ratio (α) compares collector and emitter currents and is less than unity (due to leakage). Current gain (β) compares output and input currents, its value depends on the circuit configuration and is obtained by applying Kirchoff's law to the base junction and expressing the result in terms of α. Current gain is less than unity for common base and typically between 10 and 200 for the other two configurations. Strictly current *changes* are compared. Transfer characteristic is preferably linear. Equivalent circuits and "h" parameters are used in transfer analysis.

The *pnp* transistor circuit (common emitter) is shown in Fig. 7.10. In practice the common junction is often earthed, output voltage tapped off after C to earth (via capacitor), bias battery replaced by RC circuits and supply is from voltage power lines not battery.

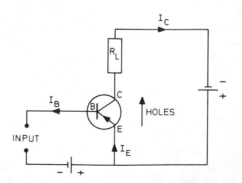

Fig. 7.10   TRANSISTOR CIRCUIT (COMMON EMITTER)

Characteristics are as shown in Fig. 7.11 where variations of base current (at input) give variations of collector-emitter voltage projected down from the load line. For the load line – where it cuts baseline is supply voltage (cut off $I_B = 0$), saturation at crossing with $I_B = 0.2$ mA here, operating point for voltage amplifier (requiring minimum distortion) at mid length approximately. Class B amplifiers biased near cut off giving half wave pulses.

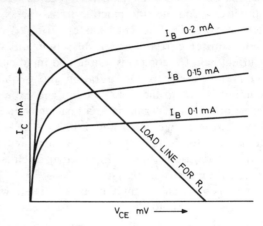

Fig. 7.11 COLLECTOR CHARACTERISTICS (CE)

Transistor Voltage Amplifier

It is necessary to include a high resistance in series with the collector and tap off voltage variations across it, 250 gain can be achieved. If connected as a common emitter a positive pulse into the base makes it less negative and fewer holes flow from emitter to base. Collector current therefore decreases and the volt drop across the resistor decreases which causes collector voltage to increase. There is a 180° phase reversal between base input and collector output voltage signals. Voltage characteristic slopes down so collector voltages decrease for base voltage increase but it is the amplification, not the negative slope, that is important. If it is necessary not to have inverted output two amplifiers in series are used – inversion is useful for negative feedback. Common base connections are desirable for voltage amplification but are difficult to design so that common emitter configurations are usually found.

Parameters

The common emitter circuit gives medium input and output impedance, good current or voltage or power gain, inversion. Common base gives low input and high output impedance, no current gain but good voltage and medium power gain, in phase operation. Common collector gives high input and low output impedance, no voltage and power gain but good current gain, in phase operation. The latter is often used for buffer stages connecting high impedance source to low impedance load (impedance matching). The common emitter mode, *pnp* or *npn*, is most frequently used for current amplification and especially for voltage amplification because a low input signal is required.

Equivalent T-Circuit

Presents a simple representation of the transistor to allow calculation of current, voltage and power gains. Such a circuit is shown in DTp – SCOTVEC Class One specimen examination question number 18 at the end of the book (common base configuration).

UNIJUNCTION TRANSISTOR

A bar of lightly doped *n* type material, with base connections B_1 and B_2 at each end, has a *p* type material contact (emitter) made between the two. If the emitter junction is reverse biased the bar acts as a potential divider. When input voltage is sufficient to drive an emitter current it flows in the B_1 emitter region. This decreases the region resistance and emitter voltage (negative resistance characteristic). The device is used for trigger operation, pulse oscillators, time delays and particularly for pulse generation to fire thyristors. In Fig. 7.12, *C* charges until voltage can eject emitter current. At a certain value of this current the voltage *V*

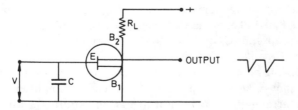

Fig. 7.12 UNIJUNCTION TRANSISTOR (OSCILLATOR)

collapses giving a negative spike output pulse with rapid discharge of C. The capacitor recharges and oscillation continues. This device is therefore a simple relaxation oscillator but is described here in the general context of the transistor.

SMALL SIGNAL JUNCTION TRANSISTOR AMPLIFIER

A linear mode device, Fig. 7.13, shows a *pnp* common emitter configuration multi (two) stage unit, the dotted line illustrates the staging, extension to more stages is achieved in the same additive way. In a triode valve auto grid bias is used and in transistors a similar principle is required. A suitable bias point which is stable with temperature variation is essential. Several ways, all dependant on feedback can be used. In Fig. 7.13 a resistor R_E in the emitter lead maintains a constant base voltage and any undue rise in the leakage (temperature induced) current causes the

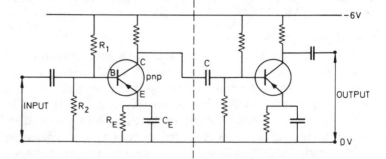

Fig. 7.13   TRANSISTOR 2 STAGE AMPLIFIER (SMALL SIGNAL)

emitter voltage to fall. The emitter-base junction approaches reverse bias so reducing current through the unit. The potential divider $R_1 + R_2$ keeps base voltage constant when there is no input signal. Capacitor C_E acts as a bypass for ac components of emitter current. Circuits typify voltage amplifiers.

Coupling between stages can either be by a transformer method or, as shown, resistor-capacitor. Whilst current amplifier design is easily arranged, voltage amplification requires a high input impedance to the first stage. Essentially this needs increased feedback which can be achieved by omitting C_E and C the interstage capacitor (direct coupling). Other methods can be used such as emitter follower but this adds to complexity. For one stage current amplifier; omit C_E and take output across R_L, connect collector directly (no resistance) to the –6V power supply.

POWER AMPLIFIER

Power dissipation must be kept low so a high efficiency is important. Class B types are preferred but single transistors can not be used because the operating characteristic range results in halfwave form outputs so that a push-pull twin arrangement is utilised. Such amplifiers require high input currents and a power (driver) stage is used, usually in Class A, the combination being classified as AB (linear mode). A typical arrangement is shown in Fig 7.14.

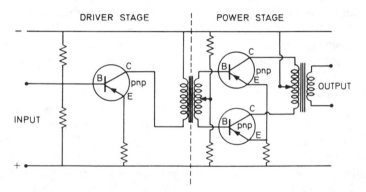

Fig. 7.14 PUSH/PULL POWER AMPLIFIER STAGE AND DRIVER

The driver transistor feeds to the transformer phase splitter. Inputs to the power stage are of equal amplitude and 180° out of phase. Each power transistor conducts for half a period and the complete waveform is restored in the output transformer. This transformer has its primary ends connected to the transistor collector leads and centre tap connected to the more negative lead. Signal flux is in the output transformer core throughout the whole period, with the complete waveform in the secondary winding. Power supplies are commonly ± 6V, as an order of magnitude.

dc AMPLIFIER

Most instrument signals require dc amplification of low voltages. This is difficult to arrange as drift variation, largely caused by temperature variation in transistors, is amplified and passed on. With ac amplifiers the coupling capacitor excludes variations but with dc units the coupling is usually direct. Differential amplifiers have been used but perfect matching is difficult. Zener diode stabilisation and feedback circuits are also employed.

The solution often adopted is to use an inverter input to derive ac from dc, direct ac amplifiers and a converter output to give dc from ac. The converter is essentially a rectifier with smoother circuit, as described in the demodulation circuit of Fig. 7.4, or a transistorised feedback integrator.

The inverter is called a **chopper** and converts steady input into square waves, height proportional to signal strength and easily amplified, by mechanical or transistor switching devices. When transistors are used voltage and temperature stabilisation are required. The transistor switch occupies four successive states, *i.e.* off (voltage applied between E and C with leakage current only flowing), on-transition (C current rising and voltage falling), on (C current flowing and saturation voltage only between E and C), off-transition (C current falling and voltage rising). Rating conditions require careful design and a small offset voltage, temperature related, arises which causes problems. A **field effect transistor** is very suited to chopping and a circuit is shown in Fig. 7.15, such transistors have negligibly small voltage offset with the distinct advantage of a very high input impedance. Junctions (JFET) utilise n or p channels whilst metal oxide silicon (MOSFET) are n channel, either enhancement or depletion type.

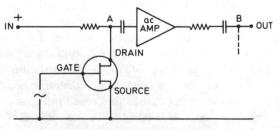

Fig. 7.15 FET CHOPPER CIRCUIT

One JFET is an *n* type bar (lightly doped) with electron input (source) and output (drain) at its ends. On both sides of the bar *p* type (heavily doped) material electrodes act as the gate (reverse biased relative to source). A narrow channel through the centre is controlled by the gate and impedance variation can arrange two conditions, saturation and "pinch off" output voltage.

In Fig. 7.15 the FET varies its drain and source resistance according to voltage level and polarity of the input signal. Essentially it is working as a make and break to produce square waveform. Output, in the absence of any restoring mechanism, would be symmetrical about a OV line as dc level is lost in the ac amplifier. The output can be periodically shorted at B with another FET as at A, so that negative going portions are removed and dc level restored. A smoothing circuit would be fitted at output. Source line could well be earthed. A FET can also give large current variations for small changes in gate voltage, *i.e.* amplifier operation, if required. It is a unipolar device. Linear and digital MOS integrated circuits are used in microprocessors in preference to bipolar types. They use variation in gate voltage to vary current source-drain, utilise one carrier and have high input impedance.

OPERATIONAL AMPLIFIER

A combination of a high gain dc amplifier. i.e. chopped, ac, smoothed, together with feedback and input impedance is required in control systems. The complete amplifier (with a gain of –A) is usually diagrammatically represented as a triangle. Such devices are described in detail in Chapter11. They are essentially voltage amplifiers, in cascade multi-stage form, gain 10^5 – as an integrated circuit on a silicon chip, used in analogue computation and digital logic.

FEEDBACK ANALYSIS

This aspect has been mentioned previously but it is now necesary to consider a more detailed analysis before concluding the work of this section. Consider Fig. 7.16 of a closed loop consisting of a feedback amplifier of forward gain G and the monitored fraction F of output fed back. Voltages are taken as the variables, V_1 input, e error, V_0 output (the more general terminology would be θ_1, θ, θ_0). Subtraction of V_0 from V_1 is arranged with an odd number of amplifier stages, giving 180° phase shift, and adding V_0 to V_1; this is **negative feedback.** Also see Page 261.

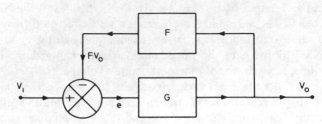

Fig. 7.16 CLOSED LOOP (FEEDBACK AMPLIFIER BLOCK DIAGRAM)

$$V_0 = Ge$$

$$e = V_1 - FV_0$$

$$V_0 = G(V_1 - FV_0)$$

$$\frac{V_0}{V_1} = G\left(1 - F\frac{V_0}{V_1}\right)$$

$$\frac{V_0}{V_1}(1 + GF) = G$$

$$\frac{V_0}{V_1} = \frac{G}{1 + FG}$$

Now if the "open loop" gain FG is high compared to 1 (*i.e.* for a high value of amplifier gain G):

$$\text{Overall Gain} = \frac{1}{F}$$

This is independent of G, which can vary, and only dependent on F which is fixed by accurate stable resistors. The resistive input impedance is increased by a factor of 1 + FG. The marked improvement in performance over the open loop amplifier is at the expense of additional amplifier stages made necessary by increasing G to make FG much greater than unity.

Consider an amplifier with a gain of 10^4 and negative feedback fraction F one-hundredth, i.e. 10^{-2}.

$$\text{Overall Gain} = \frac{10^4}{1 + 10^{-2} \times 10^4} = 99$$

Assume a change in G due to say variation of supply voltage, ageing of transistors, etc. of as much as 50% reduction.

$$\text{Overall Gain} = \frac{5 \times 10^3}{1 + 10^{-2} \times 5 \times 10^3} = 98$$

i.e. fall of 50% amplifier gain results in about 1% fall in overall gain. Negative feedback gives stability and accurate control.

Consider now the case of **positive feedback**.

$$V_0 = Ge$$

$$e = V_1 + FV_0$$

this, for electrical, is feedback voltage in phase with input voltage.

$$\frac{V_0}{V_1} = \frac{G}{1 - FG}$$

Again there is increase of overall gain compared to "open loop" gain. However there is a marked disadvantage in that a small change in G causes a much greater change in overall gain and also overall gain increases rapidly as FG increases so that when FG = 1 the overall gain is infinite. Output is available with no input, *i.e.* the amplifier is an oscillator. Positive feedback use therefore causes instability and self oscillation. Oscillators are very useful in instrumentation electronics but oscillatory situations in control systems must be avoided. Instability and oscillation shows as practical hunting or cycling on the control system which is most unsatisfactory.

OSCILLATORS

An oscillator is an instrument for producing voltages that vary in a regular fashion. The output may be sinusoidal, *i.e.* a sinewave generator (harmonic oscillator) or square, triangular, sawtooth in shape, *i.e.* relaxation oscillators. The latter are mainly important in control system equipment. A wide range of harmonic oscillators exist from high frequency (LC types) to low frequency (RC types) but apart from basic principles only the relaxation oscillator will be described because of its practical application.

BASIC THEORY

An *LC* parallel circuit in which inductive reactance equals capacitive reactance is resonant oscillatory, resonant frequency equals $1/2 \pi \sqrt{LC}$, negligible supply current and appreciable loop current (for an *LC* series circuit maximum supply current at this frequency). In practice resistance is present and oscillations would diminish. If sufficient energy, correctly timed, is supplied to the *LC* circuit utilising a transistor to make up losses the oscillation can be maintained indefinitely. This energy is supplied as correct phase and magnitude positive feedback.

Referring to Fig. 7.17, *LC* is the frequency determining tuned circuit and the remainder is oscillation maintaining. By mutual induction at *L'* positive feedback is arranged from the transistor.

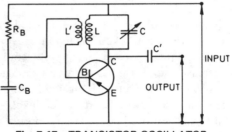

Fig. 7.17   TRANSISTOR OSCILLATOR

HARMONIC OSCILLATORS

An amplifier providing its own input functions as an oscillator and if the feedback is small is a pure sinewave generator. Typical refinements for high frequency harmonic work include Hartley and Colpitts oscillators. For lower frequency work in harmonic generation the principle is similar to that above but the feedback line includes resistors and capacitors. A three RC section input gives 180° phase shift and as a collector voltage in a transistor is 180° out of phase with base voltage this technique gives in phase between the two. A phase shift oscillator or Wien bridge network is regularly used. Linear mode signal generators are of two types namely constant voltage which is a low impedance source and constant current which is a high impedance source.

RELAXATION OSCILLATORS

Such oscillators are used in the non-linear mode where rapid switching of transistors from off to saturated (bottomed) is arranged. For sawtooth waveform a unijunction transistor, as previously described and illustrated in Fig. 7.12, can be used. Alternatively a gas discharge tube with RC circuit is suitable usually with a thyratron (or thyristor) controlling time base strike voltage. A more sophisticated method is to use a single amplifier connected as an integrator (see Chapter 11). High current pulses of short duration can be obtained from a blocking oscillator.

For rectangular waveform generators a push-pull blocking oscillator can be used as shown in Fig. 7.18. The transformer core material has a square form hysterisis loop. When supply is

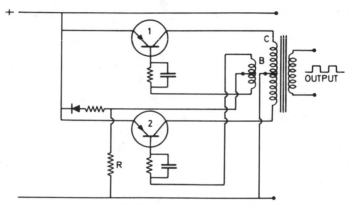

Fig. 7.18   PUSH/PULL BLOCKING OSCILLATOR

switched on (high starting resistance R) assume transistor 1 starts conduction so that a voltage is developed across the collector winding (C) and induced across base winding (B). Polarity is such that 1 is forward and 2 reverse biased so that the positive feedback turns 1 full on and 2 is off. Collector current from 1 rapidly builds up and the core of C saturates and voltage collapses which cuts off base drive to 1 and collector current starts to fall. A voltage is now induced in B to try and maintain this current but reversed in sense. 2 is now biased on and 1 off and the half cycle is repeated.

Two state circuits, with abrupt transition from one state to the other, can be arranged with RC coupled amplifiers in place of the transformer coupling described. The basic circuit has the output of one transistor connected to the input of another and vice versa. Such devices are called free running (astable) multivibrators. If instead of two unstable states. only one is arranged, the circuit is called a monostable triggered multivibrator or univibrator (flip-flop). The effect of a pulse is to flip the circuit into the unstable state whereupon normal multivibrator action allows flop back into the stable state. A flip-flop circuit is described in Chapter 16. Another method is to "clip" sinusoidal waves using zener diodes. Power supplies often ± 6V, *i.e.* of this order of magnitude.

OTHER DEVICES

Electronic digital devices are greatly used in logic circuits and computing, typical applications are considered in Chapter 16. The speed of response and reliability of transistor switching greatly enhance its use in annunciator and control circuits.

Certain specialist equipment, mainly used in analysis, such as ultra violet recorders, wave and transfer function analysers, etc. is best described from specialist literature if required. Perhaps the most important in this respect is the cathode ray oscilloscope (CRO)

CATHODE RAY OSCILLOSCOPE

The oscilloscope is shown in Fig. 7.19. The full circuit includes power supply packs and amplifiers. In addition a relaxation oscillator timebase (multivibrator saw tooth generator) is required.

In Fig. 7.19 the electron beam from the gun is focussed on to a fluorescent screen and the glass tube is under high vacuum (pressure 10^{-6} mm Hg). Input signals represented as voltages are connected to the Y plates and deflections produced are shown on the vertical axis of the screen. The timebase circuit is connected to the X plates and the beam is deflected horizontally from the left to right with uniform speed and then returned in almost zero time.

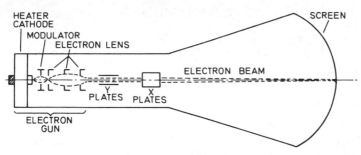

Fig. 7.19   CATHODE RAY OSCILLOSCOPE

Applications in analysis are numerous. A relevant practical application of the CRO is display of a pressure volume diagram from an IC engine. Screen scales can be calibrated directly in the required units and the application of integration gives power and torque characteristic on display.

LIGHT EMITTING DIODE (LED)

Emits light when forward biased. Gallium arsenide gives out energy as infra red radiation; phosphides and indium give green and yellow. LEDs are useful on-off or fault indicators and can be used in arrays, etc.

RADIO COMMUNICATION

The production, transmission and detection of sound or picture information is by electromagnetic waves – the range and spectrum of such radiation is given in Fig. 7.20 (see also Fig. 7.4).

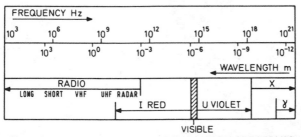

Fig. 7.20   ELECTROMAGNETIC RADIATION SPECTRUM

FIBRE OPTICS

Modulated emitted light (diode/laser) is pulsed at high frequency through optical glass fibre cable then converted to electrical signal.. Advantages – easy multiplexing, no interference or hazards.

TEST EXAMPLES 7

1. Explain why electronic devices are inherently suitable for display systems and generally unsuitable for actuating systems. Describe any method by which a small electrical signal may be amplified.

2. State the purpose served by transistors. Sketch the circuit diagram for an *npn* voltage amplifier. Explain the precautions which should be taken regarding care, handing and operating environment for transistorised devices.

3. Describe a halfwave rectifier. Sketch the output wave form: (i) without, (ii) with a capacitor across the purely resistive load terminals. What modifications are necessary to give full wave rectification?

4. Draw a circuit diagram of a single stage common emitter transistor amplifier. Indicate the type of transistor and the supply polarities. Give details of one cause of distortion that can arise in the circuit.

CHAPTER 8

FINAL CONTROLLING ELEMENTS

The element which acts directly on the controlled body, process or machine. In a servo-mechanism this is a servo-motor, rectilinear or rotary, receiving amplifier output and driving the load. Correcting unit (motor and correcting element) is applicable to process control systems. An actuator is a motor with limited rotary or rectilinear motion.

CORRECTING UNITS

Such devices may be pneumatically, hydraulically or electrically operated or a combination and the word motor is applicable to all.

DIAPHRAGM OPERATED CONTROL VALVE

This unit, as described, is pneumatically operated. Description can be given in three parts namely motor, valve and positioner. The sketch given (Fig. 8.1) refers to a fuel variable but is applicable to any similar variable. Ring gland seals of Teflon, Neoprene bonding or aluminium foil braiding are used, often lubricated.

Motor element

Air pressure acts on top of a synthetic rubber diaphragm and is opposed by upward spring force, oil flow is right to left, hand regulation is possible and the fail-safe position is shut (up). The pressure-stroke characteristic is based on linear which requires a large constant area diaphragm, minimum friction and a linear spring force-deflection characteristic. A limited travel motor-actuator-*reverse* action.

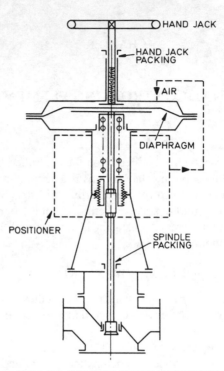

Fig. 8.1   FUEL OIL CONTROL VALVE WITH POSITIONER

Correcting element (valve)

The valve can be single seated reverse action as shown, or direct action single seated, or double seated (direct or reverse) which give balanced valve forces and less operating energy and are widely used. Materials for all components depend on the medium being controlled. The overall flow characteristic requires to be assessed for the piping system as a whole, as well as for the valve, to achieve design conditions. In general valves may be simplified into *three* types of variable % Flow—% Valve Lift characteristics. Mitre valves with wings (bevel or poppet) usually give inverted near parabola characteristics best suited to on-off operation. Vee port (in wings) high lift or modified parabolic contoured valve plugs can be designed for proportional control. Characteristics are then a true linear relation between flow and valve travel for the former and near parabolic equal percentage, *i.e.* equal increments of valve travel give equal percentage change in existing flow, for the latter. These three types are sometimes defined as quick opening, linear and semi-logarithmic.

$$\text{Turn Down Ratio} = \frac{\text{Maximum controllable flow \%}}{\text{Minimum controllable flow \%}}$$

i.e. each a percentage of the theoretical (100%) flow.

Positioner

Such devices are necessary when:

(a) there is a high pressure drop across the valve,
(b) the valve is remote from the controller,
(c) the medium being controlled is viscous,
(d) there are high gland pressures required.

Essentially each of the above effectively increase friction, hysteresis or unbalanced forces acting on the valve spindle. The positioner provides extra power to position the valve accurately and speedily to offset these effects.

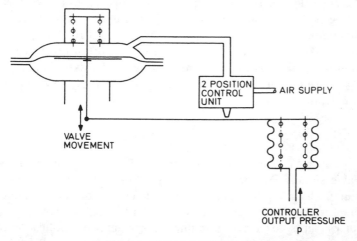

Fig. 8.2   VALVE POSITIONER

A motion feedback device from the valve spindle senses deviation between the desired value position input signal and the actual valve position and supplies extra correcting power. A flapper is connected at one end to the valve spindle and to a pressure bellows at the other, with the nozzle between the two. Increase of p increases pressure on the diaphragm until valve movement restores the flapper to the throttle position and equilibrium is restored. Further speed boost can be added, as shown, by inserting a pneumatic relay to apply full air pressure to the diaphragm. Both positioner and relay (as described in Chapter 10) act as pneumatic amplifiers. Valve positioners are very often used in sequence operation of valves in a system, *i.e.* split range control (Chapter 13). Double seated valves (two inlets through ends, outlet at centre) are often used giving three way mixing and bypass arrangements on engine coolant systems.

PISTON OPERATED CONTROL VALVE

Double seated valves with balanced valve forces are invariably utilised with diaphragm control valves because of the characteristics of this actuator. However the piston type actuator gives powerful valve forces, long stroke and accurate positioning so that single seated valves can be used which often have a more desirable flow pattern and require less maintenance. In the *direct* acting type a loading pressure on top of a piston (down force) is maintained constant (supply air via combined reducing-relief valve). Actuating air on the bottom of the piston (up force) is controlled in pressure by a small relay pilot valve, diaphragm operated from input signal, and connecting to supply (open up) or vent (close down).

TORQUE ACTUATED CONTROL VALVE

There are two air motors, one for each direction of rotation of the valve spindle. Motors are transverse to spindle and rotate the spindle by a ratchet and toothed wheel. Motors are standard diaphragm or piston mechanisms. High torques for large valve forces can be achieved by multiple diaphragms or duplication of motors. A diagrammatic sketch is shown in Fig. 8.3.

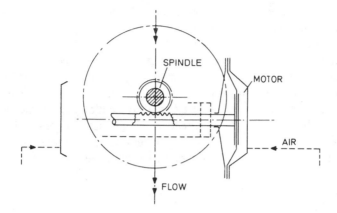

Fig. 8.3   TORQUE ACTUATED VALVE

An alternative design utilises electric motor gear drive to the spindle by a similar arrangement. The motor must be reversible, variable speed and fitted with limit switches.

ROTARY CYLINDER CONTROL VALVE

The valve consists of a semi-rotary shutter operating in the square throat section of the valve passage. Operating torque is transmitted to the shutter gate via a spindle perpendicular to flow direction. The "swing through" design of butterfly valve requires a good spindle seal and tight shut off is best arranged by closure of shutter against a flexible disc. Such valves are best suited to throttling operation between about 15° and 60° for reasonable turn down ratios. Torque requirement is not high and linear actuators of any design are suitable. Equal % flow characteristic.

WAX ELEMENT TEMPERATURE CONTROL VALVE

This type has a copper capsule containing wax whose expansion varies with temperature. Movement is transmitted via a diaphragm, plunger and linkage to vary the position of a shutter which rotates in the valve body. The valve is best suited to mixing or bypass conditions, control normally being limited to a range of temperature of about 10°C, fail-safe inherent in design. While the self contained simple design is attractive for many duties it is a fixed control. The sketch of Fig. 8.4 is a diagrammatic representation. Temperature rise causing down-movement of the plunger, clockwise rotation, closing right and opening left outlet connection. The basic three port valve body can be designed with the shutter operated by an external pneumatic cylinder operated from a remote pressure controller activated by a temperature sensitive element.

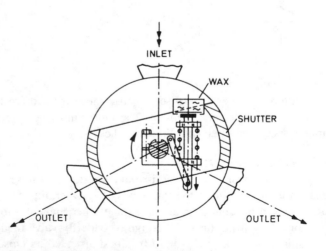

Fig. 8.4   WAX ELEMENT ACTUATED VALVE

SERVO-MOTORS

May be rectilinear or rotary; operated by air, fluid or electricity; applied in either process or kinetic control systems.

dc ELECTRIC MOTORS

Essentially the servo-motor is a conventional motor, series, shunt or compound, with control of field current or armature voltage by the controlling device. High torque and low inertia is required so that armatures are reduced in diameter and lengthened. Good commutation over a wide range of speeds is necessary and design must allow for peak transient changes. Performance is limited by heating caused by high armature currents and magnetic saturation of iron paths. Reversal is arranged by reversing the current through the field or armature via the controlling device, which is generally satisfactory, although split field motors can be used.

ac ELECTRIC MOTORS

The three phase induction motor is a most desirable machine in electrical work, being cheap and reliable. Unfortunately starting torque is low and the torque-speed characteristic is non-linear so that control is difficult. For servo use the torque characteristic can be improved by using high resistance rotors which unfortunately generate extra heat and cost. Commutator motors are available but add to complexity and cost. Thyristor circuitry offers the most possibilities for improvement.

The two phase induction motor is used in low power systems especially for position control. Applications include instrument potentiometers, bridges and pen recorders. Such a motor has two stator coils wound at right angles, which are fed with alternating currents 90° out of phase, to produce the rotating magnetic field. For reasonable modulation, torque is proportional to the two currents. If a fixed voltage and frequency is applied to one reference winding, then torque is proportional to the voltage of the other winding, which is connected to represent the amplitude of the control signal. Characteristics, especially with a high resistance rotor, are reasonably linear over a limited range. Heat generation at reference field and rotor are high.

Single phase motors for servo systems are unsatisfactory except for on-off control and special starting arrangements, such as split phase, are required.

Synchronous motors can be used for low power drives such as pen recorders, etc. where synchronism of timing is required.

HYDRAULIC RAM SERVO

Used for linear actuation, *i.e.* ram or jack type, but can be utilised as a torque device with multiple rams. Generally a medium control performance system used in position devices. Normally short stroke but can operate with long stroke and high pressures (ship steering gear). There is little in the construction calling for special attention.

HYDRAULIC VARIABLE DELIVERY PUMP

Not strictly a final controlling element but has important applications in hydraulic control systems with output to hydraulic servo-motor (linear or rotary). The swash plate pump is a good example with alternatives such as Hele-Shaw eccentric slipper, tilting trunnion, cam operated ball piston, etc. The hydraulic motor is virtually a reverse pump and has similar construction. Consider Fig. 8.5 illustrating the swash plate type:

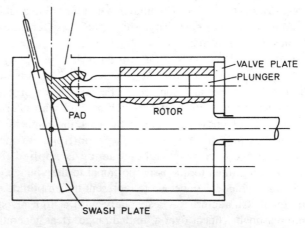

Fig. 8.5 VARIABLE DELIVERY PUMP

Slipper pads bear against the swash plate face and plungers are driven in and out axially for each revolution of the rotor. The swash plate movement varies effective stroke and can reverse the flow. A number of axial plungers are used in the rotor. Delivery can be to an identical motor with fixed swash plate.

Other types, particularly using a booster supply pump, have a similar design and can give discharge pressures over 140 bar. Higher pressures give smaller components and very positive action.

HYDRAULIC ROTARY VANE SERVO

Details are as sketched in Fig. 8.6. Rotation depends on which side of the vane is connected to the pump pressure feed, this should be clear from the plan view as sketched. The large rotary vane unit is normally designed for a maximum pressure of about 90 bar as distortion and leakage are liable to occur at higher pressures. The design is simple and effective. In fact the apparent space and weight saving is not as great as may be imagined due to the higher pressures and integrated construction utilised in modern hydraulic ram designs. There is however, a definite space saving but the first cost is higher. Absorption and transmission of torque relief is essential to avoid excess radial loading of vanes. The three vane type is used for rudder angles to 70°. Steel sealing strips backed by synthetic rubber are fitted into grooves along the working faces of rotor and stator vanes.

PNEUMATIC PISTON SERVO

Refer to Fig. 8.7:

The pilot valve has two outlets, one to the top and one to the bottom of the servo-piston. If the valve is displaced from its neutral position then pressure at one port increases whilst at the other port it decreases, so causing piston movement. The movement of the piston is arranged, via linkage gear and cam, to vary the tension on a spring giving an opposing moment to the signal pressure on the bellows. When these two moments balance the pilot valve is at mid or neutral position and the pressures on each side of the servo-piston balance, so locking the piston.

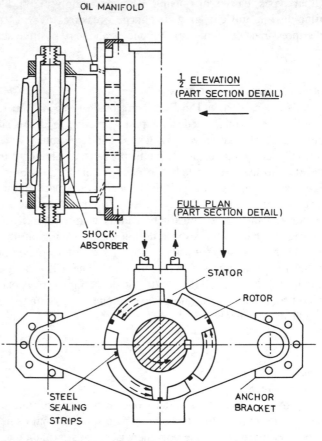

Fig. 8.6   HYDRAULIC ROTARY VANE SERVO

(In the case of a diaphragm valve the pilot valve would only have one outlet to the diaphragm top, the valve stem movement via the gear linkage and cam would alter the spring restoring movement until equilibrium existed.) An incoming air signal inflates the bellows causing the balance beam to pivot about the right hand end and operate the pilot valve. This produces a second air signal whose relationship to the first signal is dependent on the tension in the spring attached to the beam end point.

The mechanical movement of the valve, or other control device being operated, is transmitted by a driving rod to the cam and linkage, thereby adjusting the tension in the spring. By using a suitably shaped cam the position of the regulating unit to which the positioner is attached may be given a predetermined relationship with the incoming air signal.

$^1/_{120}$ bar change in signal pressure will give full travel to the pilot valve which could give $1^1/_3$ bar variation in output pressure.

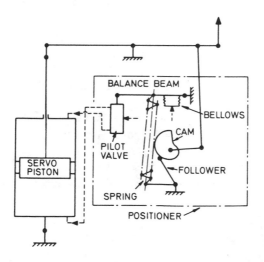

Fig. 8.7 PNEUMATIC PISTON SERVO AND POSITIONER

OTHER SERVO-MOTORS

Variable couplings and clutches can be used, either hydraulic or electric, for rotary speed control. Electromagnetic solenoid devices, linear and rotary, are available for many purposes including electrical contact and relay operation and incremental digital stepping motors.

TEST EXAMPLES 8

1. Make a detailed sketch of a simple diaphragm operated control valve, such as a reducing valve, although any other type of a similar control valve will be accepted. Analyse the action of the interconnecting elements, that is, those parts affecting control.
Explain how load changes are sensed and the command signals are transmitted to the actuator.

2. An air damper is controlled in position by variable air pressure to a pneumatic actuator. Sketch and describe the system, describe the actuator and explain its mode of operation.

3. Sketch and describe a control valve of the wax element type and show how it is incorporated into a coolant system.

4. With the aid of simple sketches explain briefly what is meant by:
    i. a direct-acting diaphragm operated valve;
   ii. a reverse-acting diaphragm operated valve;
  iii.. a linear characteristic valve;
   iv. an equal percentage characteristic valve.
Make a diagrammatic sketch of a valve positioner and explain its action.

CHAPTER 9

PROCESS CONTROL THEORY

TERMINOLOGY

The following definitions should be considered with Fig. 9.1 to which they mainly refer round the loop system. Other definitions of terms are given at the appropriate place in the text to cover and clarify essential points. Process control is concerned with physical quantities involving variables such as temperature, pressure, flow, level, etc.

About twenty terms are exactly defined immediately below as from British Standards which together with Figs 9.1 and 9.4 are from the same source.

These extracts are from B.S. 1523: Part 1: 1967, a Glossary of Terms used in Automatic Control and Regulating Systems (Part 1. Process and Kinetic Control). They are reproduced by permission of the British Standards Institution, 2 Park Street, London, W1Y 4AA, from whom copies of the complete standard may be obtained.

CONTROL SYSTEM

An arrangement of elements (amplifiers, converters, human operators, etc.) interconnected and interacting in such a way as to maintain or affect in a prescribed manner, some condition of a body, process or machine which forms part of the system.

PROCESS CONTROL SYSTEM

A control system, the purpose of which is to control some physical quantity or condition of a process.

COMMAND SIGNAL

The quantity or signal which is set or varied by some device or human agent external to and independent of the control system and which is intended to determine the value of the controlled condition. (Symbol γ_i.)

SET VALUE *(Set point)*

The command signal to a process system.

DESIRED VALUE

The value of the controlled condition which the operator desires to obtain.

INPUT ELEMENT

The element which is included, when necessary, to convert the actual command signal into a converted command signal suitable for operating the comparing element.

Note. In some systems there is no input element as the command signal is taken, without conversion, direct to the comparing element.

CONVERTED COMMAND SIGNAL

A physical quantity related only to the command signal, and normally proportional to it, but of a different physical kind suitable for operating the comparing element or the coordinating element. (Symbol θ_i.)

Note. In definitions where there is no ambiguity the term "command signal" will be used to imply the command signal itself or the converted command signal.

COMPARING ELEMENT

The element which accepts, in physically similar form, the command signal and the controlled condition, or their equivalents, and determines the deviation or the converted deviation.

DEVIATION

The difference between the measured value of the controlled condition and the command signal. (Symbol $\gamma = \gamma_0 - \gamma_i$.)

CONVERTED DEVIATION

A physical quantity related only to the deviation, and normally proportional to it, but of a different physical kind suitable for operating the amplifier element. It may also be used for operating other elements in the system. (Symbol $\theta = \theta_0 - \theta_i$.)

(Author's note: often written $\theta = \theta_i - \theta_o$, $\gamma = \gamma_i - \gamma_0$).

SIGNAL PROCESSING

The processing of the information contained in a signal by modulating, demodulating, mixing, gating, computing or filtering.

CORRECTING UNIT

Of a process control system. The single unit containing the motor element and correcting element in a process control system.

MOTOR ELEMENT

The element which adjusts the correcting element in response to a signal from an automatic controller.

CORRECTING ELEMENT

The final controlling element in a process control system.

CONTROLLED CONDITION

The physical quantity or condition of the controlled body, process or machine which it is the purpose of the system to control. (Symbol γ_0.)

DETECTING ELEMENT

The element which responds directly to the value of the controlled condition.

MEASURING ELEMENT

The element which responds to the signal from the detecting element and gives a signal representing the controlled condition.

LOAD

The rate at which material or energy is fed into, or removed from, the plant (*on a process control or regulating system*). 1. The controlled device. 2. The properties (*e.g.* inertia, friction) of the controlled device that affect the operation of the system (*of a kinetic control system*).

MEASURING UNIT

A unit which gives a signal representing the controlled condition. It comprises a detecting element and measuring element.

Note. Such a unit is used as the monitoring element of a process control system.

CONVERTED CONTROLLED CONDITION

A physical quantity related only to the controlled condition and normally proportional to it, but of a different physical kind suitable for operating the comparing element or the co-ordinating element. (Symbol θ_0.)

Note. In definitions where there is no ambiguity the term controlled condition will be used to imply the controlled condition itself or the converted controlled condition.

AUTOMATIC CONTROLLER *(Automatic regulator)*

A portion of an automatic controlling or regulating system in which a signal representing the controlled condition is compared with a signal representing the command signal and which operates in such a way as to reduce the deviation.

Note. The two functions of an automatic controller or regulator, namely to determine the deviation and to generate the control signal dependent on the deviation, are in many devices carried out by two separate parts, the comparing element and the controlling element respectively.

MONITORING FEEDBACK

The feedback of a signal representing the controlled condition along a separate path provided for that purpose, for comparison with a signal representing the command signal to form a signal representing the deviation.

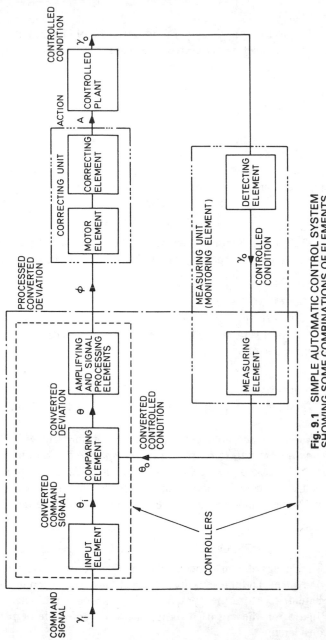

Fig. 9.1 SIMPLE AUTOMATIC CONTROL SYSTEM SHOWING SOME COMBINATIONS OF ELEMENTS

PROCESS

The act of physically or chemically changing, including combining, matter or of converting energy.

OFFSET *(Droop)*

Sustained deviation.

OVERSHOOT

The difference between the maximum instantaneous value of the step function response and its steady state value.

DEAD TIME

The time interval between a change in a signal and the initiation of a perceptible response to that change (a dead band region may exist on a controller).

CASCADE CONTROL SYSTEM

A control system in which one controller provides the command signal to one or more other controllers.

SETTLING TIME

The time taken to approach a final steady state within specified limits.

Note. The settling time depends on the limits specified and is meaningless unless these limits are specified. Example. In the particular case of a series *LR*   circuit subject to a step function of voltage, the settling time of the current to within 1 per cent of its final value is approximately five times the **time constant** L/R.

RESPONSE OF DETECTION ELEMENTS

Time lags obviously occur in a plant due to the individual lags of components and transmission of signal lags. The lags must be fully evaluated before the control design can be established. As an illustration the lag of a temperature detector element can be considered:

Consider a detector element which is directly inserted in a pipeline. The fluid flowing increases in temperature at a uniform rate of say 10°C in one minute.

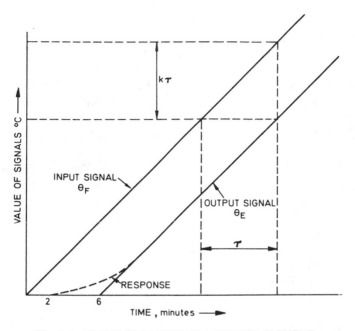

Fig. 9.2   LINEAR RESPONSE OF DETECTOR ELEMENT

Referring to Fig. 9.2:

The first indication of temperature change at the detector element may be after about say 2 minutes and there may be a constant lag at a given reading of about 6 minutes (ramp input).

If θ_F is the fluid temperature and θ_E is the element temperature:

$$\theta_F - \theta_E \propto C_E \, R_F$$

C_E is thermal capacity of element (mass times specific heat)
R_F is liquid to element thermal resistance to heat flow

$$\theta_F - \theta_E = kC_E R_F$$

where k is a constant.

Measuring lag is $\theta_F - \theta_E$ (say °C)

$C_E R_F$ is the **time constant** τ (say minutes)

$$\theta_F - \theta_E = k\tau$$

This lag consideration is based on linear variation, if the variation was exponential then the measure lag is usually arbitrarily defined in terms of the time it takes for the output signal amplitude to reach 63·2% of the input signal amplitude. The lag time on the sketch is given in minutes but should preferably be reduced to seconds in practice. Fig. 9.2 assumes fairly heavily damped response. Under-damped response would show in oscillation curves about the line θ_E.

If the disturbance causing the variation is a continuous sine form variation the appearance would be as in Fig. 9.3, note that the re-

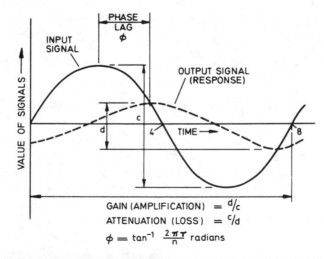

GAIN (AMPLIFICATION) $= \frac{d}{c}$

ATTENUATION (LOSS) $= \frac{c}{d}$

$\phi = \tan^{-1} \frac{2\pi\tau}{n}$ radians

Fig. 9.3 STEADY STATE RESPONSE TO SINUSOIDAL SIGNAL

sponse has a reduced amplitude and has a phase lag. τ is the time lag of detecting element, n is the period of process disturbance, attenuation applies as gain is less than unity.

Considering the case of a temperature detector element in a pocket then to reduce lags the time constant (CR) must be reduced to give quick response and the following would be aimed at:

1. A close fitting thermometer in a pocket with immersion in a high conductivity fluid.

2. Clean fittings and a high velocity for the fluid to be measured (turbulent flow).

3. Light, good conductivity pocket material using deep immersion into the flow of fluid.

4. Reduced piping distances, friction and inertia effects.

DISTANCE VELOCITY LAG

That time interval between an alteration in the value of a signal and its manifestation unchanged at a later stage arising solely from the finite speed of propagation of the signal.

For example the time it takes for a heating effect to travel with the fluid from heat source to detection element along a lagged pipeline. Lag = Distance/Velocity. Causes phase lag. Theoretically no magnitude change.

TRANSFER LAG

That part of the transmission characteristic, exclusive of distance-velocity lag, which modifies the time-amplitude relationship of a signal and thus delays the full manifestation of its influence.

For example the measure lag as given previously for the detection element which is dependent on R and C. Causes phase lag and reduces amplitude.

The aim is to keep inherent lags as small as possible, together with reducing inertia and increasing stiffness (or their equivalents), for the system. The alternative is to increase system gain but this can create instability.

TYPES OF CONTROL ACTION

Can be illustrated by any variable; level is selected for this chapter.

TWO STEP CONTROLLER ACTION **(Basic action. 1.)**
The action of a controller whose output signal changes from one predetermined value to another when the deviation changes sign.

It should be noted that the limits are not necessarily on-off although this is often used (see Fig. 9.4), especially in digital systems (Chapter 16).

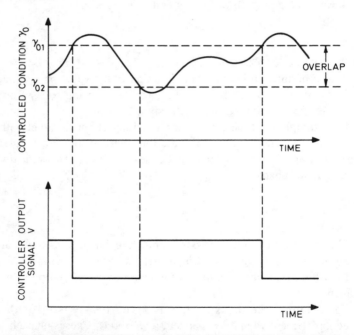

Fig. 9.4   THE ACTION OF A TWO STEP CONTROLLER WITH OVERLAP

A typical example of two step controller action would be liquid level control in a tank with a varying supply and a required continuous and steady outflow. Overlap, which could be adjustable, allows working between predetermined limits (differential gap or overlap) and gives less irregular action, this would be suitable, as another example, for refrigeration motor cut in and out control by room temperature.

Lag tends to allow overshoot and this should normally be reduced to reasonable limits. The closeness of control is influenced by the capacity of the system and also that property of an un-controlled system to reach equilibrium for a fixed set of conditions (inherent regulation).

Such action of two step control is a simple but most useful method with numerous applications in practice. By arranging more chosen values and corresponding correcting signal steps the control can be made much closer with less overshoot, this method is termed multi-step controller action.

Note. The more simple the control principle the better. Additions or refinement such as integral action, derivative action, etc., as covered later, should only be applied where the requirements of the process control definitely require these modifications.

PROPORTIONAL CONTROL ACTION (P) **(Basic action. 2.)**

The action of a controller whose output signal V (or Φ) is proportionate to the deviation θ.

θ is the difference between the measured value of the controlled condition θ_0 and the command signal θ_i.

$$V \alpha - \theta$$
$$V = - K_1 \theta$$

The negative sign denotes that the correction signal is opposite in direction to the deviation. K_1, a constant depending on the *controller characteristic,* is called the proportional action factor.

Potential correction Φ (*change* of *actual* controlled condition γ_0) is proportional to the movement of the correcting unit (which depends on V).

$$\Phi \alpha V$$
$$\Phi = C_1 V$$

where C_1 is a constant depending on the *correcting unit characteristic*.

Now
$$V = -K_1\theta$$
$$\Phi = -K_1 C_1 \theta$$
$$\Phi = -\mu\theta$$

$\mu = \Phi/\theta$ and is numerically the proportional control factor, or simply the controller gain, a typical value, in pneumatics, may be about 15.

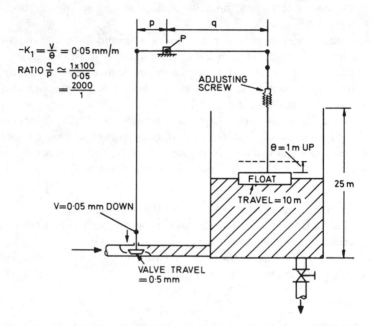

Fig. 9.5 LIQUID LEVEL CONTROL BY SELF OPERATING CONTROLLER

Referring to Fig. 9.5 and assuming linear characteristics. It is desired to maintain a fixed height h in the tank, the outflow demand varies. As this is self operating (no intervening medium such as compressed air) then we can utilise V, K_1 and θ as symbols. If h is set for 10 m, assume the valve is then 0.25 mm from the seat, $K_1 =$ 0.05 mm/m (decided by leverage), then if h increases to 11 m the valve movement to shut in is 0·05 (11 − 10) = 0·05 mm, *i.e.* the new valve position is 0·3 mm from the seat. There is no controller gain here, much the *reverse* in fact.

Proportional band

That range of values of deviation corresponding to the full operating range of output signal of the controlling unit, from proportional action only.

This band can be expressed as a percentage of the range of values of the controlled condition which the measuring unit of the controller is designed to measure (see Figs. 9.6 and 9.7).

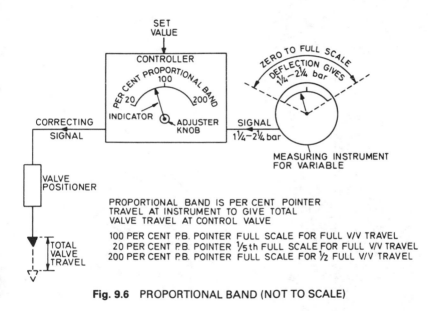

Fig. 9.6 PROPORTIONAL BAND (NOT TO SCALE)

$$\text{Proportional band} = \frac{(\text{Total valve span})}{(\text{Total measure span}) \times K_1} \times 100$$

For the example for the given level controller, if the full measurement scale is from 0 to 20 m head, *i.e.* 100%, and the full valve stroke is 0·5 mm, *i.e.* 100% then 10 m fully strokes the valve (*i.e.* 0·05 × 10 = 0·5).

$$\text{Proportional band} = \frac{10}{20} \times 100 = 50\%$$

alternatively:

$$\text{Proportional band} = \frac{0 \cdot 5}{0 \cdot 05 \times 20} \times 100 = 50\%$$

If the proportional bandwidth is narrow then a big controlling movement is required for a small deviation, the control is sensitive, *i.e.* high value of K_1 and a small offset results (see later). Too narrow a proportional bandwidth can however cause instability and hunting. The practical result is a compromise, the set value must of course be within the band.

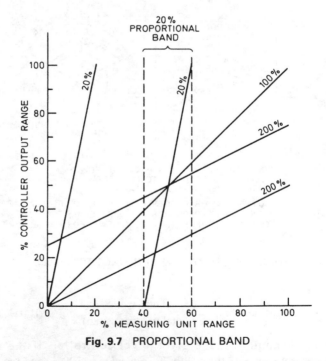

Fig. 9.7 PROPORTIONAL BAND

Note:

For good control the following are essential:

1. A high deviation reduction factor (hence high μ), *i.e.* small deviation from set value after a disturbance. High μ, means high K_1, highly sensitive, narrow proportional band, etc.

2. Minimum offset.

3. Low value of subsidence ratio at short oscillating period, *i.e.* quick return to set value after a disturbance.

These are achieved by plant analysis. Widening a proportional band causes an increase of offset, of damping and of period of oscillation.

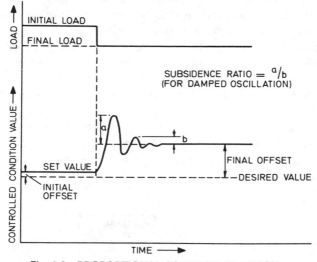

Fig. 9.8 PROPORTIONAL CONTROLLER ACTION

Offset

Is sustained deviation due to an inherent characteristic of proportional control action. (It should be noted that with all proportional controllers the *set value* differs from the *desired value* by varying amounts depending on the given load conditions.) If K_1 is large, for a given deviation, the offset will be small, K_1 is dependent on the proportional band of the controller (see Fig. 9.8). Consider the following analogy (as shown in Fig. 10.4):

A spring hanging from a support, with a mass of say 10 kg on the end, will have a certain extension and the mass will be at a certain vertical position X. If the mass is pulled down and released it will oscillate but will finally return to X.. This is an example of proportional action in that the spring restoring force is proportional to the extension (also opposite in direction). The desired value position X is always reached.

Now if the mass is changed to say 20 kg then after oscillation it *will not* return to X but *will* return to a new position of equilibrium Y. Offset will be the difference between the X and Y positions.

This means, with proportional control, where a set value (the command signal to a process control system) at a given load occurs it will *only give coincidence with the desired value at that load.* At any other load offset exists. If this offset is acceptable to a plant the proportional controller is satisfactory. If the offset is too great additional refinements have to be incorporated into the controller (see Fig. 9.10, Fig. 9.11 and integral action later). *In many cases of practical description in this book and elsewhere, for simplicity, desired value, set value or set point, etc., are all used for the same thing and no distinction is made on a fine point of principle.*

Fig. 9.9 illustrates another analogy using level control, *i.e.* equilibrium before and after a demand change with two different heads and valve settings.

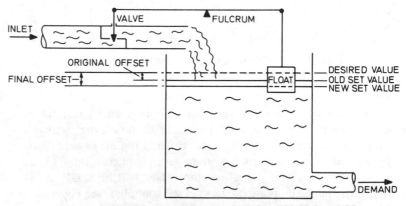

Fig. 9.9 SIMPLE PROPORTIONAL ACTION CONTROL LOOP

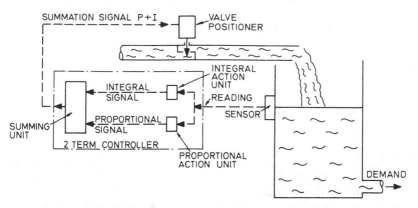

Fig. 9.10 PROPORTIONAL PLUS INTEGRAL CONTROL LOOP

For a **human control loop.** A man regulates the water inlet valve to a tank to maintain gauge level with variable outflow demand. He is told the level required (desired value), will see the level (measured value), after a change will compare the two and decide on adjustment (correcting action), finally there is amplification for muscle action to operate valve (correcting element). Proportional control will arrest the change and hold it steady but at a point different from the original set value due to the load change. The human operator would bring the level back to the desired value after arresting the level, *i.e.* he would apply re-set (integral) action to remove offset. Overshoot would not occur because the operator would not, while adjusting for offset, go on altering the valve right up until offset was gone. He would ease down valve adjustment rate as desired value was approached, *i.e. apply a damping action, based on rate (derivative).*

INTEGRAL CONTROL ACTION (I) **(Basic action. 3.)**

The action of a controller whose output signal changes at a rate which is proportional to the deviation.

Note:

The object of integral control action is to reduce offset to zero.

$$\frac{dV}{dt} \propto - \theta$$

by the definition given above

$$\frac{dV}{dt} = - K_2\theta$$

$$V = - K_2 \int \theta\, dt$$

K_2 is called the integral action factor.

$$\Phi = C_1 V$$

$$\Phi = - C_1 K_2 \int \theta\, dt$$

$$\Phi = - p \int \theta dt$$

p is called the integral control factor.

Thus the potential correction Φ at a given time t is proportional to the area between the desired and recorded values $\int_0^t \theta\, dt$ is a mathematical way of writing that area). Rate of change of potential correction with respect to time is proportional to the deviation {$d\Phi/dt$ *expresses mathematically rate of change of* Φ *with respect to* t}.

Now

$$\mu = C_1 K_1$$

and

$$\rho = C_1 K_2$$

$$\therefore \frac{K_1}{K_2} = \frac{\mu}{\rho} = S$$

Integral action time S

By definition:

In a controller having proportional plus integral action, the time interval in which the part of the output signal due to integral action increases by an amount equal to the part of the output signal due to proportional action, when the deviation is unchanging.

After time S then $V_p = V_I$ and the controller output change equals $2V_p$. Note that $S = \mu/\rho$, *i.e.* integral action reduces controller gain. The *larger* the S time setting the *less* the contribution from in-

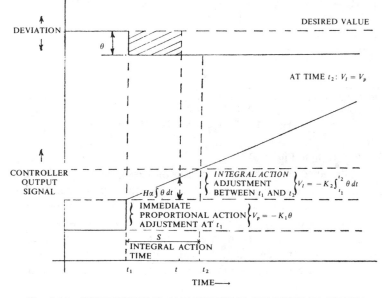

Fig. 9.11 PROPORTIONAL AND INTEGRAL CONTROLLER ACTION

tegral action to the potential correction. S is increased by increasing resistance.

Fig. 9.11 is given assuming an instantaneous deviation change. This change is referred to as *step function input*. Deviation under proportional action alone would constantly increase, giving bigger offset, but the addition of integral action maintains a constant deviation.

A *more likely* short term action would give integral action until deviation ceased, *i.e.* no offset. Integral action reduces a previously fixed controller gain μ and introduces extra lag, so is undesirable, it should not be used unless the use of a wide proportional band gives too big offsets. See Fig. 9.12.

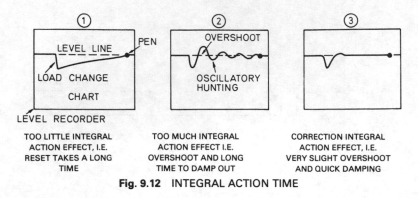

Fig. 9.12   INTEGRAL ACTION TIME

Referring to Fig. 9.13:

If the level rises, the small piston moves up and high pressure fluid flows through the top port and returns through the port B, this action via the link closes the valve to reduce the inflow. This movement will always continue as long as a deviation exists and the rate of travel depends on the area of the top port opening which is proportional to the deviation. Conversely fall in level causes oil to flow in at the bottom port and return through port A. The only time

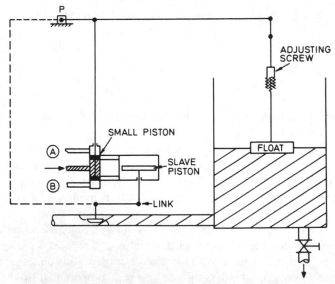

Fig. 9.13   INTEGRAL ACTION ON LIQUID LEVEL CONTROL

the valve is not moving is at the desired value, offset will never be possible.

Note:

Proportional plus integral ($P + I$) two term can be applied by including the link shown dotted in Fig. 9.13. For a rising float, above set point, both act in the same direction downwards to close the valve, pivot P can be moved to vary the individual actions. For a falling float *above* set point the actions are in opposition. For a falling float *below* set point both act to open the valve. Integral action is *always* tending to reduce offset. Integral action is *not* used alone, if it was the characteristic would be similar to two step action. Note the use here of a hydraulic controller (small and slave piston) for the integral action itself. Proportionally controlled first order open loop systems, with an inherent integration characteristic (for example shaped lavatory cisterns), respond exponentially to a stop function when the loop is closed.

DERIVATIVE CONTROL ACTION (D) **(Basic action. 4.)**

The action of a controller whose output signal is proportional to the rate at which the deviation is changing.

Note:

The object of derivative control action is to give quicker response and supplement inadequate proportional control damping.

$$V \propto - \frac{d\theta}{dt} \quad \text{by definition above.}$$

$$V = -K_3 \frac{d\theta}{dt}$$

K_3 is called the derivative action factor.

$$\Phi = C_1 V$$

$$\Phi = - C_1 K_3 \frac{d\theta}{dt}$$

$$\Phi = - \eta \frac{d\theta}{dt}$$

η is called the derivative control factor.

Now

$$\mu = C_1K_1$$

$$\eta = C_1K_3$$

$$\therefore \frac{K_1}{K_3} = \frac{\mu}{\eta} = \frac{1}{T}$$

Derivative action time T

By definition:

In a controller having proportional plus derivative action, the time interval in which the part of the output signal due to proportional action increases by an amount equal to the part of the signal due to derivative action, when the deviation is changing at a constant rate. Such deviation change is referred to as *ramp function input*.

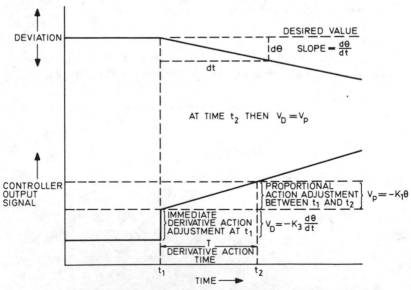

Fig. 9.14   PROPORTIONAL AND DERIVATIVE CONTROLLER ACTION

In Fig. 9.14 the derivative action is assumed instantaneous. After time T then $V_D = Vp$ and total controller output signal change equals $2Vp$. Use of derivative action, increases μ and gives phase lead, which are desirable, but T is limited as too much derivative action may cause instability and hunting. Note that $T = K_3/K_1$, the *larger* the T setting the *greater* the derivative action contribution to the potential correction (increased by increasing resistance).

Referring to Fig. 9.15:

Now from mechanics, rate of change of displacement with respect to time is velocity. At reasonable velocities damping resistance is proportional to velocity, so one would expect some form of damping device in this rate response action.

Derivative control is *not* used alone, it is a transient condition which must be combined with proportional control..

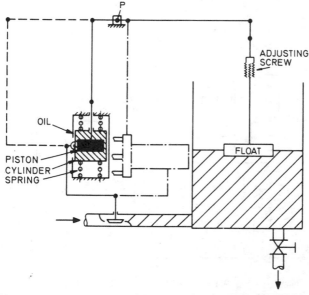

Fig. 9.15   DERIVATIVE ACTION ON LIQUID LEVEL CONTROL

If the level rises at a certain rate the piston (in the dashpot) moves down at a certain velocity, proportional to this velocity is a down force on the cylinder which acts to close the valve in, the cylinder motion is resisted by a spring. Valve displacement is proportional to down force. Whenever the float stops changing position the down force ceases and the springs return the cylinder to its original position.

Note:

Proportional plus derivative $(P + D)$ two term can be applied by including the link shown dotted, proportional plus derivative plus integral $(P + D + I)$ three term can be applied by including the link shown dotted and the link shown chain dotted. For a rising float, for $(P + D + I)$ with the arrangement shown, above set point, all act in the same direction downward to close the valve, pivot P can be moved to affect the value of all control factors.

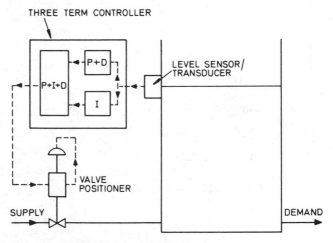

Fig. 9.16 THREE TERM CONTROLLER

THREE TERM CONTROLLER

$P + I + D$ actions, combined, are illustrated for level control in Fig. 9.16.

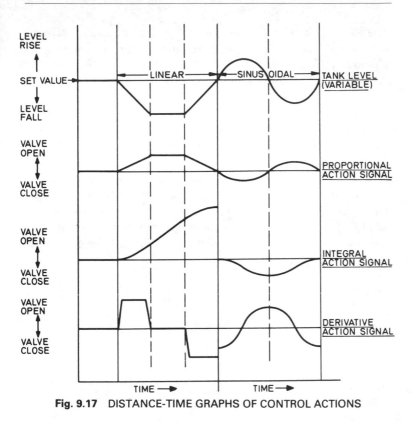

Fig. 9.17   DISTANCE-TIME GRAPHS OF CONTROL ACTIONS

DISTANCE TIME GRAPHS OF CONTROL ACTIONS

Refer to Fig. 9.17:

Such analysis gives a clear pictorial representation. Slope of a distance time graph is velocity; an inclined straight line is constant velocity as the slope is constant, a curve of decreasing slope represents deceleration. The top two diagrams should be self explanatory, relative heights depend on proportionality factor. For integral action note that whenever the variable is away from desired value the integral effect is always moving to correct. For the value at any instant on the third sketch think of area developed at that instant on the first sketch, applied on the opposite side of the axis, and to a suitable scale factor. For derivative action note that it opposes the motion of the variable irrespective of the desired value.

The value of the signal on the fourth sketch is the change of slope of the first sketch, again opposite side of axis, *i.e.* slope only changes at four points on sketch one and at such points the derivative effect is acting almost instantaneously.

CASCADE CONTROL

Consider a multiple capacity system for level control, as an example a two capacitor tank system illustrated in Fig. 9.18. Single capacitor systems respond quickly to load changes and are easy to control utilising correct proportional band and reset action but interaction occurs with multiple systems. Tank A acts as a lag effect on the controlled process from tank B so the combination is less sensitive, especially to supply variations, this is an inherent problem with large inertia (mass, heat capacity, etc.) systems as for example IC engine coolant circuits.

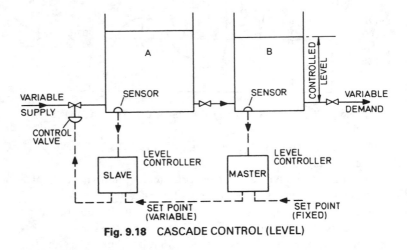

Fig. 9.18 CASCADE CONTROL (LEVEL)

Consider Fig. 9.18:

There are two variables, supply and output demand, affecting the controlled variable which is level in tank B. The slave controller with level sensing from tank A controls the input supply control valve according to the set point and is a single capacitor control loop for tank A. The master controller with level sensing from tank

B (the controlled variable) controls the input supply quantity to tank B, *i.e.* the level in tank A, and is a single capacitor control loop for tank B. This is achieved because the master controller signal controls the set point of the slave controller. A two capacitor system has therefore been simplified to two single capacitor systems which are more easily controlled. Alternatively sensing for the slave could be flow rate at supply rather than tank A level. The process can be extended to multi-capacity systems with control of any desired variable. The principle is utilised very often in practice, for example IC engine coolant and Butterworth heating, as described in Chapter 13. If a certain pressure range of controller output is divided up by relays or controllers for different functions, in for example a sequence, this may be termed **split range control.** This is also utilised in practice, for example exhaust range pressure control (Chapter 13).

EXAMPLE—LEVEL CONTROL

When considering numerical questions it is often best to utilise a tabular approach as the following example illustrates:

The sketch (Fig. 9.19) shows a single element boiler water level control system. Assuming that the system has been adjusted so that the level is at the desired value of 16 cm ("half glass") when the load is 500 kg/min, determine:

(a) the offset if the load is reduced to zero,

(b) the proportional band setting required such that the offset is limited to 8 cm if the load changes from 500 kg/min to 100 kg/min.

Steam flow load (kg/min)	500	0	100
Level (cm)	16	32	24
Controller input (kN/m²)	60	100	80
Range change (kN/m²)		40	20
Controller output (kN/m²)	70	20	30
Range change (kN/m²)		50	40

i.e. offset 16 cm, proportional band 50%. Unless stated to the contrary, a linear proportionality is assumed between the indicator/controller variable scale ranges in such cases. A similar

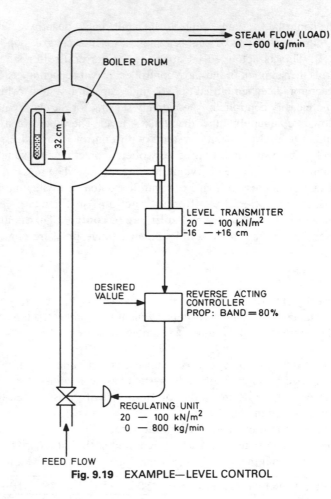

STEAM FLOW (LOAD)
0 — 600 kg/min

BOILER DRUM

32 cm

LEVEL TRANSMITTER
20 — 100 kN/m²
-16 — +16 cm

DESIRED
VALUE

REVERSE ACTING
CONTROLLER
PROP: BAND = 80%

REGULATING UNIT
20 — 100 kN/m²
0 — 800 kg/min

FEED FLOW

Fig. 9.19   EXAMPLE—LEVEL CONTROL

question is included at the end of the book—specimen examination question number 15, HND (BTEC & SCOTVEC). An alternative method of solution to that previous is: System proportional control factor μ equals multiple of proportionality characteristics/ coefficients, *i.e.* $31·25 = 2· 5 \times 1·25 \times 10$ when $\Phi = - \mu \, \theta$ and as $\Phi = - 500$ kg/min so $\theta = + 16$ cm. Similarly working in the reverse direction from $\theta = + 8$ cm the controller characteristic/ coefficient, or proportional action factor K_1 is now 2 (gain), proportional band is 50%.

   Similar questions are included at the end of the book, including those requiring graph plots of controller signals—after analysis by a method such as just described. When the controller includes such as integral action the approach is similar but it must be remembered that integral action time (S) elapses while the signal changes (by integral action) by an equal amount to the immediate proportional action signal. See HNC (BTEC & SCOTVEC) specimen examination questions number 16 and number 18 HND (BTEC & SCOTVEC).

TEST EXAMPLES 9

1. Explain the meaning of the following terms relating to process control:
   (a) desired value,
   (b) error signal,
   (c) detecting element,
   (d) feedback,
   (e) reset action,
   (f) servo-motor.

2. Explain the meaning of the following terms, using suitable diagrams where appropriate:
   (a) potential correction,
   (b) proportional control,
   (c) integral control,
   (d) integral action time,
   (e) derivative control,
   (f) derivative action time.

3. Draw simple diagrams showing the response of a detecting element suffering from a distance velocity lag equivalent to 5s and a single transfer lag, when subjected to disturbances in the form of:
   (a) a step,
   (b) a ramp.
Illustrate, on two simple diagrams, the effect of an increase in frequency on "phase lag" and "attenuation" for a detecting element suffering from transfer lag.

4. A temperature measuring device suffers a distance-velocity lag of 15s and also exhibits a simple transfer lag with a time constant of 40s. If the temperature being monitored jumps suddenly from 30°C to 35°C, what temperature does the device indicate 55s after this step change? (33·16°C.)

CHAPTER 10

PNEUMATIC CONTROL PRINCIPLES

Note:
The bar is used as the unit of pressure in this section. 1 bar = 10^5 pascal (Pa) = 10^5 N/m^2.

PNEUMATIC TWO STEP CONTROL TECHNIQUE
Referring to Fig. 10.1:
The constriction may be about 0·2 mm bore and the nozzle about 0·75 mm outlet bore, these sizes are largely fixed by air purity condition, *i.e.* particle filtration size.

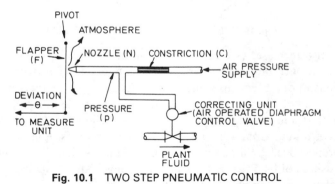

Fig. 10.1 TWO STEP PNEUMATIC CONTROL

With the flapper, or baffle, moved away from the nozzle full nozzle pressure drop occurs, pressure p may be about 1·2 bar or less. With the flapper almost closing the nozzle, pressure p may be near 2 bar, i.e. almost supply pressure. Two values of pressure p can be arranged, which will depend on the flapper position, which is in turn decided by the measure signal movement. An on-off operation, or low-rate and high-rate operation, can be utilised with these two pressures.

Average flapper travel between two limits is often less than 0·05 mm, the nearer the flapper to the nozzle the stronger the measure signal force required, this is a limitation. The relation between flapper travel and pressure p is *non-linear, i.e.* equal increments of flapper travel do not give equal increments of pressure p, but over a fairly wide range of travel, say 30 to 70%, the relation is *reasonably linear, i.e.* linear between 0·015 and 0·033 mm in 0·05 mm travel..

THE RELAY

Provides pneumatic amplification, proportional movement, and reduced time lag. Equivalent in action to an electronic amplifier.

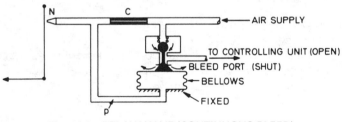

Fig. 10.2 RELAY VALVE (CONTINUOUS BLEED)

If p increases (see Fig. 10.2) then the bellows acts to close the bleed port and supply air passes, conversely if p decreases a continuous bleed to atmosphere occurs. Amplification by a fraction of 16 can easily be arranged, for example a flapper travel of 0·01 mm causing a change of p on the bellows of 0·05 bar could give output from 1.2 to 2 bar. Flapper travel is approximately proportional, by a *linear* relation, to output signal in *this* throttle position of 0·05 bar bellows pressure range. An alternative type of relay is given in Fig. 10.3 for comparison

Various alternatives and refinements can be added, for example, a bellows connected to the relay output will give action proportional to output utilising negative feedback, etc. The relay could introduce further non-linearity if not properly matched in design.

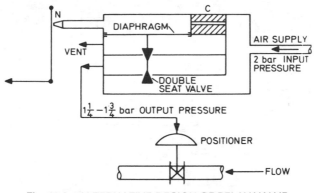

Fig. 10.3   ALTERNATIVE DESIGN OF RELAY VALVE

PNEUMATIC PROPORTIONAL CONTROL TECHNIQUE

This utilises adjustable (negative) feedback due to the bellows and flapper linkage. Input (from the measure signal) is compared to output (from the relay signal) and the action is to reduce this difference, so matching input to output. Any desired ratio between input and output can be achieved by adjusting the linkage ratio $a : b$ shown in Fig. 10.4. For a 50% proportional band then the measurement change is 50% of scale for full valve stroke, i.e. under ideal conditions the control should operate to maintain measured value and desired value together at 50% valve stroke. Varying load means the controller keeps conditions stable within the proportional band, but not at the desired value, maximum offset cannot exceed half band width.

Referring to Fig. 10.4:

Consider the measure link moving right, this decreases the nozzle escape and pressure p increases. Pressure p acts, via the relay, on the bellows so tending to act in the opposite direction to the initial movement with proportional action against the spring (just as for the simple spring analogy p. 142 earlier). This decreases the sensitivity (flapper travel near nozzle). The ratio $a : b$ (which is adjustable) decides the bandwidth, this action is the simple lever principle (feedback can never exceed deviation).

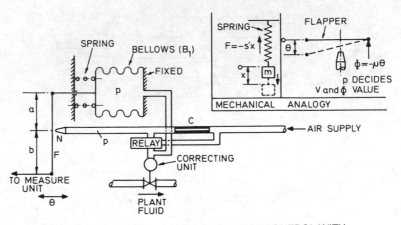

Fig. 10.4 PNEUMATIC PROPORTIONAL CONTROL WITH NEGATIVE FEEDBACK

This cancellation whereby a pressure increase moves the flapper to lower the pressure means that a greater movement of the measure unit, for a given change in control line pressure, is required so ensuring proportional action. This also gives a wider proportional band without increasing mechanical linkages which would reduce accuracy. The relay shown in all sketches is not in practice necessarily fitted to all controllers. Without the bellows the proportional flapper travel region is very small.

Bellows movement is proportional to pressure p.

$$x = m\theta - np$$

where x is movement of flapper next to the nozzle, m and n are proportionality constants (including the adjustments a and b) for the deviation θ and the negative feedback pressure p (which also decides V and Φ).

x is negligible compared to other movements.

$$\therefore np = m\theta$$

$$\therefore p = -\frac{m}{n}\theta$$

$$\therefore \Phi = -\mu\theta$$

i.e. potential correction is *proportional* to the deviation and *equals* the proportional control factor multiplied by the deviation. Negative sign to indicate opposite direction.

STACK TYPE CONTROLLER PRINCIPLE (P ACTION)

Referring to Fig. 10.5:

The construction is of air chambers stacked on top of each other separated by diaphragms and incorporating relay valves, nozzles and restrictor control valves.

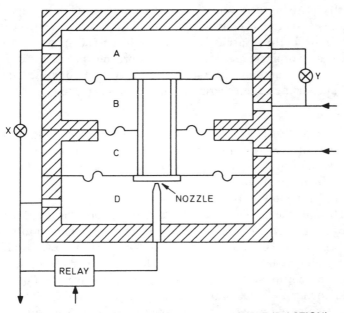

Fig. 10.5 STACK TYPE CONTROLLER PRINCIPLE (P ACTION)

The measured value (controlled condition) enters at chamber C and the set value (command signal) set up at the reducing valve enters at chamber B. Variations between these two values causes the diaphragm arrangement to move up or down vertically so that air flow through the nozzle to the chamber D controls cancellation of the deviation caused by the pressure variation.

Pressures at A and B would be equal if control valve X were closed. hence pressures at C and D would equalise as D pressure (controller output signal) changed, this means 100% proportional band.

Conversely if X were opened fully pressures at A and D would be equal so that deviation from set value would cause the nozzle to be fully opened or closed. This is two step control action. Valve Y can act as an adjustment but essentially it prevents direct connection between output and set value air lines.

Variation of the setting of X between open and closed gives proportional band variation between 0% and 100%.

By utilising different stacking arrangements, e.g. capacity bellows, different connections to control valves (restrictors), variable bellows areas, etc. then the correct proportional, derivative or integral actions can be incorporated as required. *P, P + D, P + I* actions separately generated and combined in an addition unit give three term action without interaction problems.

PNEUMATIC PROPORTIONAL PLUS INTEGRAL CONTROL TECHNIQUE

Integral (reset) action can be regarded as a slow cancellation of the sensitivity reduction provided by the negative feedback of the proportional system.

Referring to Fig. 10.6:

Without the needle value adjustable restrictor I, the proportional negative feedback bellows B_1 effect would be completely cancelled by the proportional positive feedback bellows B_2 effect (assuming equal bellows sizes and form), simulated two-step or near proportional action for limited flapper travel would result (depending on flapper travel utilised). Similarly in the steady state *with* the needle value as there would be zero pressure difference across it.

When a disturbance causes a deviation to occur (say p increases) then the rate of p_2 change is proportional to the deviation effect $p - p_2$.

If the measure unit moves right under a constant deviation increase then p increases giving near proportional action Vp *immediately. Negative feedback to bellows* B_1 *reduces sensitivity* giving wider proportional band and true proportional action.

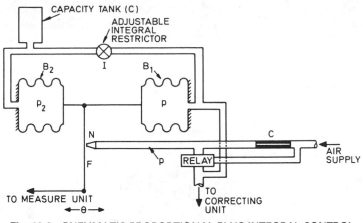

Fig. 10.6 PNEUMATIC PROPORTIONAL PLUS INTEGRAL CONTROL
WITH NEGATIVE FEEDBACK

Deviation, under proportional action alone, would become greater, more offset would occur. However bellows B_2 exerts positive feedback to raise pressure p_2 at a rate dependent on the deviation, this maintains constant deviation. Integral action would continue until deviation ceased and there would be no offset, *i.e.* restoration to desired value. A repeat is accomplished when the amount of change in air pressure in the reset (integral or floating) bellows equals the amount of original change in output pressure to the proportional bellows.

$$x = m\theta - n(p - p_2)$$

i.e. negative feedback due to p and positive feedback due to p_2. Again taking $x = 0$:

$$p - p_2 = \frac{m}{n}\,\theta \quad \text{for whole action}$$

now
$$\frac{dp_2}{dt} \propto p - p_2 \quad \text{for integral action}$$

i.e. rate of change of pressure p_2 is proportional to the difference of pressure.

$$\frac{dp_2}{dt} = \frac{1}{CR}\,(p - p_2)$$

CR is a time constant dependent on the capacity C of the tank and the resistance R of the restrictor I, CR is the integral action time S.

$$S \frac{dp_2}{dt} = \frac{m}{n} \theta$$

integrating

$$Sp_2 = \frac{m}{n} \int \theta \, dt$$

$$p_2 = \frac{m}{nS} \int \theta \, dt$$

Thus for the whole action

$$p = \frac{m}{n} \theta + \frac{m}{nS} \int \theta \, dt$$

applying negative direction sign, with p equivalent to Φ

$$\Phi = -\mu \left(\theta + \frac{1}{S} \int \theta \, dt \right) \text{ for } (P + I)$$

Note:
Consider the distinct analogies:

Pressure: $\dfrac{dp_2}{dt} = \dfrac{1}{CR} (p - p_2)$

where rate of pressure p_2 with respect to time is proportional to excess pressure $p - p_2$. CR time constant, C tank capacity and R restrictor flow resistance.

Electrical: $\dfrac{dV_2}{dt} = \dfrac{1}{CR} (V - V_2)$

where rate of change of voltage V_2 with respect to time across a condenser is proportional to excess voltage $V - V_2$. CR time constant, C capacitance of the condenser and R current resistor resistance, the latter often negligibly small.

Temperature: $\dfrac{d\theta_E}{dt} = \dfrac{1}{CR} (\theta_F - \theta_E)$

where rate of change of temperature θ_E across a detector element with respect to time is proportional to excess temperature $\theta_E - \theta_F$. CR time constant, C is thermal capacity of element and R thermal resistance to heat flow.

PNEUMATIC PROPORTIONAL PLUS DERIVATIVE CONTROL TECHNIQUE

Derivative (rate) action increases sensitivity by restricting the negative feedback provided by the proportional system so that during the change high speed sensitivity occurs but when the change ceases ordinary proportional action occurs.

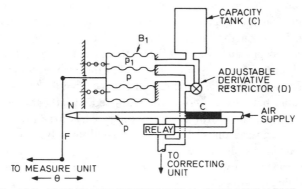

Fig. 10.7   PNEUMATIC PROPORTIONAL PLUS DERIVATIVE
CONTROL WITH NEGATIVE FEEDBACK

Referring to Fig. 10.7:

Without the needle valve adjustable restrictor D, proportional action results, similarly for the steady state with the needle valve. When a disturbance causes a deviation to occur (say p increases) then the *rate* of p_1 change is proportional to the deviation effect $p - p_1$. Derivative action stabilises more quickly after a change. Considering the inner bellows, then the smaller area gives a less force per unit pressure change of p, a narrower proportional band, higher sensitivity and less feedback occurs than if the same pressure effect acted on the larger outer bellows.

If the effective area of the inner bellows was one quarter of that of the outer bellows then an instantaneous deviation produces a finite response equivalent to four times the normal proportional action with negative feedback for the same actuating signal applied to the outer bellows.

Considering the outer bellows which gives the derivative action then the change of pressure via the relay across the resistance and tank (D and C) gives a pressure drop proportional to the rate of chance of activating signal deviation. This means the sensitivity reduction due to negative feedback is adjusted in line with the rate of change of the deviation.

The combination bellows means that when movement starts (say to the right) to increase, the narrow proportional band, caused by p only on the small bellows, gives high output relay signal pressure. Such exaggerated output is then amended for derivative action by p_1 on the large bellows until the measure movement ceases and pressure in the outer bellows equals pressure in the inner bellows (no pressure drop). This means the control valve operates sooner for the same rate of change from the measure unit.

Double bellows are *not* always utilised, strictly the derivative action is on the outer bellows only. This arrangement has disadvantages as phase lag to the derivative action occurs (see compound controllers later).

The inner bellows gives proportional action only, for simplicity regard this bellows as omitted.

$$x = m\theta - np_1$$

negative derivative action feedback due to p_1.

Taking $x = 0$ as previously:

$$p_1 = \frac{m}{n}\theta \quad \text{for whole action.}$$

Now

$$\frac{dp_1}{dt} = \frac{1}{CR}(p - p_1) \quad \text{for derivative action only.}$$

CR is the time constant, dependent on capacity C of the tank and resistance R of the restrictor D, CR is derivative action time T.

Rearrange the last expression:

$$p = p_1 \left(T \frac{d}{dt} + 1 \right)$$

$$p = \frac{m}{n} \theta \left(T \frac{d}{dt} + 1 \right)$$

$$p = \frac{m}{n} \left(T \frac{d\theta}{dt} + \theta \right)$$

$$p = \frac{m_1}{n_1} \left(T \frac{d\theta}{dt} + \theta \right)$$

where m_1 and n_1 are new proportionality constants to allow for the proportionality feedback effect of the combined bellows.

Now applying the negative sign to indicate the opposite direction and with p equivalent to potential correction Φ.

$$\Phi = -\mu \left(\theta + T \frac{d\theta}{dt} \right) \quad \text{for } (P + D)$$

PNEUMATIC COMPOUND CONTROLLER $(P + I + D)$

Three term $(P + I + D)$ or two term $(P + I$ or $P + D)$. A controller action in which the output signal from the controller is the result of more than one operation on the deviation.

$$V = -K_1 \left(\theta + \frac{K_2}{K_1} \int \theta \, dt + \frac{K_3 \, d\theta}{K_1 \, dt} \right) \quad \text{for three term}$$

$$\Phi = -\mu \left(\theta + \frac{1}{S} \int \theta \, dt + T \frac{d\theta}{dt} \right) \quad \text{for three term}$$

The sketch of Fig. 10.8 shows the compound pneumatic controller, the action should be clear from previous diagrams.

Interaction

With the pneumatic arrangement shown in Fig. 10.8, but D at X, adjustment of either I or D affects each of the three actions. Thus the effective action times (S and T) differ from the nominal (dial set) action times. D can be moved from X to Y which may improve performance but still gives interaction. For truly independent adjustments then derivative and integral actions should be generated separately based on proportional action and combined in a relay (see also Chapter 15).

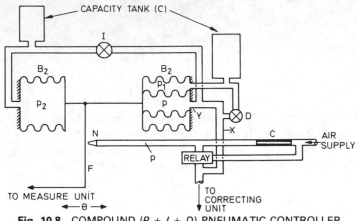

Fig. 10.8   COMPOUND (*P* + *I* + *D*) PNEUMATIC CONTROLLER

AIR SUPPLIES

An adequate supply of clean, dry compressed air is required with well designed, installed and maintained air line systems.

Quantity is defined under standard intake conditions, *i.e.* 15° C and 1 bar, which relate size, capacity and consumption.

Quality requires that filtration removes solid particles, oil and water. If dew point can be reduced at high temperature, below any likely ambient temperature of the system, the installation can be kept dry. High compression, with interstage and after cooling is effective especially when large delivery receivers allow cooling under pressure. Absorber filters such as silica gel or activated alumina should be fitted at low level system points to act as moisture traps – such traps should also be fitted adjacent to reducing valves.

Compressors are either arranged to run continuously, fitted with unloading devices to allow running light when pressure supply is reached, or have pre-set cut in and cut out pressure switches. Machine capacity should be such that 50% time loading only is required; a 3 kW unit (4 bar) would meet this criteria when delivering output of one cubic metre (referred to standard conditions) for up to 10 instruments.

Air lines should have a gradient of at least 1:50, with moisture traps at lowest points, and instrument tappings are taken from the top of headers. Final stage filters are often of the ceramic type and silicone impregnation makes them water repellant. Annealed seamless copper tubing, pickled inside and out, is often used – especially for single instrument loads. Polythene and PVC tubing is resistant to corrosive atmospheres and is also much cheaper for larger installations. For the supply to say 10 instruments the hp line (4 to 7 bar) would be about 12 mm bore, delivery through 18 mm reducing valve and filter to a 25 mm bore lp header (at 1.5 bar).

TEST EXAMPLES 10

1. Describe the operating principle of a pneumatic controller. Explain what is meant by the term "proportional action". Show by means of a simple sketch how the controller functions to maintain a particular system in equilibrium.

2. (a) Make a diagrammatic sketch of a two term ($P + I$) pneumatic controller of a nozzle-flapper type and briefly describe its operation.

   (b) With reference to the diagram describe briefly how the controller could be set before commissioning.

   (c) With reference to the diagram describe briefly how the proportional band adjustment could be calibrated.

3. (a) Make a diagrammatic sketch of a three term nozzle flapper controller which has provision for receiving pneumatic desired value and measured variable signals.

   (b) Explain how the proportional action is generated and how the gain may be varied.

   (c) Describe how the integral action is generated and how the degree of integral action may be varied.

   (d) Describe how the derivative action is generated and how the degree of derivative action may be varied.

   (e) Explain why a relay valve is necessary and what additional benefits may result from its use.

4. Describe, with the aid of a suitable sketch, the construction of a P + D controller. Explain, using a graph, the open loop response to a ramp change in measured value. How can the controller be changed from direct-acting to reverse-acting?

CHAPTER 11

ELECTRONIC CONTROL PRINCIPLES

Note:
V has been mainly used previously as the symbol for controller output signal. In this chapter V is used as a general symbol for voltage, suffix i or 1, etc for input (alternative E or e, for V, sometimes used elsewhere).

THE OPERATIONAL AMPLIFIER
Amplification, generally voltage, is readily achieved for ac or dc. It is easier to amplify ac than dc using simple circuits and dc amplifiers suffer from drift due to voltage variations. Measure signals are however dc so that many process controllers utilise dc amplifiers. Amplifier gain, *i.e.* ratio of output to input, can reach 10^8, without negative feedback from the last stage, and 10^4 is common. Inclusion of such feedback allows reduced, and fixed, ratios between output and input utilising resistive and capacitive components to generate control actions. Essentially there are input and feedback networks and the amplifier adjusts output voltage so that the amplifier takes negligible current, *i.e.* equal currents in the two networks. There must be a voltage polarity change to balance currents, i.e. 180 degrees out of phase, which requires an odd number of amplifier stages. Gain of circuit depends only on resistance ratio which is accurate and independent of any variation in amplifier gain. The amplifier inverts and multiplies; it is an active component as distinct from a passive electrical network. The high gain amplifier can use negative feedback to reduce gain to what is actually required and the negative feedback gives

stability. Energy input is required (See also Chapter 7, operational amplifier, feedback analysis).

Figure 11.1 illustrates the operational amplifier circuit (power supplies to amplifier omitted, earths shown). The signal input voltage is usually applied at one terminal (V_1) and the other input terminal is usually earthed, similarly for output and amplifier.

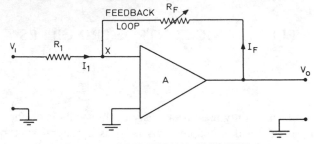

Fig. 11.1 OPERATIONAL AMPLIFIER

Alternatively two inputs can be fed into A and V_0 will then represent the difference. I_1 is negligibly small and zero potential effectively applies at X, *i.e.* infinite input impedance as an ideal. Due to inversion, feedback is essentially negative. Input resistance may well be of the order of one megohm, output resistance some tens of ohms, amplifier gain say 10^5, overall gain say 100. The upper lead to A (often marked – inside triangle) is commonly called the inverting input, lower lead (+) is then the non inverting input. Magnitude of power supplies for the circuit depends on the power requirements of the amplifier itself. Response to positive and negative inputs is required and amplifier power needs require say + 6V. – 6V and 0V connections (these are omitted from sketches for simplicity). Voltage used depends on conditions, input will be less.

ELECTRONIC TWO STEP CONTROL TECHNIQUE

On-off devices find a wide range of application using simple switch technique. A typical example is room temperature control in which a bi-metallic strip closes or opens electrical connections leading to energy input. Small permanent magnets ensure snap action and differential gap is adjustable. Digital devices are often used.

ELECTRONIC PROPORTIONAL CONTROL TECHNIQUE

Apply Ohm's law to the circuit of Fig. 11.1.

$$I_1 = \frac{V_1 - Vx}{R_1} = \frac{V_1}{R_1} \quad \text{as } Vx \text{ is zero}$$

$$I_F = \frac{V_0 - Vx}{R_F} = \frac{V_0}{R_F} \quad \text{as } Vx \text{ is zero}$$

$$I_1 = -I_F$$

because amplifier current is zero.

This is Kirchhoff's law for currents at a point.

$$-V_0 = \frac{R_F}{R_1} V_1$$

Action is *scalar multiplication* (V_0 proportional to V_1) with the multiplying factor R_F/R_1 and the negative sign indicating *inversion*. Adjustable circuit gain (proportional band) is achieved by altering R_F with R_1 fixed ($R_F/R_1 = 1$ is 100% bandwidth). For multiplying factors below unity an adjustable potentiometer (attenuator) can be used. $V_0 = tV_1$ where $t = r/R$ the tapping ratio. Removal of the minus sign can be achieved by using two amplifiers in series. $G = R_F/R_1$ is commonly 1 or 10, gain.

Consider now two inputs as shown in Fig. 11.2.

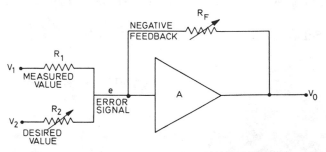

Fig. 11.2 ELECTRONIC PROPORTIONAL CONTROL

$$V_0 = -R_F \left(\frac{V_1}{R_1} + \frac{V_2}{R_2} \right)$$

where V_1 and V_2 are negative inputs. This is essentially a *summer* (and scalar) action which can be extended to further inputs as required. If $R_1 = R_2 = R_3 = $ etc. $= R_F$ then $V_0 = -(V_1 + V_2 + V_3 +$ etc.), i.e. summation only. *Ratio* control of inputs can be achieved by adjustment of the respective input resistances. A controller must produce an output signal proportional to the *deviation* between desired and measured values. The two signals can be compared (comparing element), perhaps elsewhere, by opposition flow through a common resistor the voltage across which now represents deviation (error) signal, for transmission to the amplifier input. If one input voltage is applied as negative to a summer the result is effectively *subtraction*. For Fig. 11.2 if V_1 and V_2 are regarded as measured value and desired value (in opposition) this gives error input and *proportional to deviation* control action, bandwidth adjustment at R_F. Amplifier power supplies and earth lines are omitted for simplicity.

ELECTRONIC INTEGRAL (RESET) CONTROL TECHNIQUE

Consider the circuit of Fig. 11.3:

By placing a capacitor C_F in the feedback circuit a limit is placed on the amplifier response rate to change of input signal.

For a capacitor $\quad C = Q/V$

$$\therefore Q_F = C_F V_0$$

$$I_F = \frac{dQ_F}{dt} = C_F \frac{dV_0}{dt}$$

$$\int dV_0 = \frac{1}{C_F} \int I_F dt$$

but $\qquad I_1 = -I_F$ and $I_1 = \frac{V_1}{R_1}$

$$V_0 = -\frac{1}{C_F R_1} \int V_1 \, dt$$

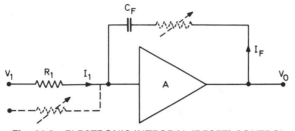

Fig. 11.3 ELECTRONIC INTEGRAL (RESET) CONTROL

i.e. output voltage is the integral voltage with the time constant (reset rate) dependent on C_F and R_1. If, as for the two previous sketches, there is a modification to two inputs, via resistors representing measured and desired values, then amplifier input voltage corresponds to error voltage, *i.e.* output voltage is the integral of error input voltage. A feedback resistor R_F is necessary to give proportional addition and make adjustment more easy, with a fixed capacitor, alternatively potentiometer adjustment could be provided. Integral action is essentially rate control in the feedback network of the circuit by capacitance. R_F and R_2 additions shown dotted. Integral action is very rarely applied on its own.

ELECTRONIC DERIVATIVE (RATE) CONTROL TECHNIQUE

Consider the circuit of Fig. 11.4 in which it is necessary to consider $P + D$ combinations:

A capacitor C_D is in the input circuit, together with a resistor R_1, to produce a rate of change component.

In the steady state there is no current through C_D.

$$I_1 = -I_F$$

$$V_0 = -\frac{R_F}{R_1} V_1$$

In the transient (changing) state:

$$I_1 = \frac{V_1}{R_1}$$

$$I_C = C_D \frac{dV_1}{dt}$$

$$I_1 + I_C = - I_F$$

$$I_1 + I_C = - \frac{V_0}{R_F}$$

$$V_0 = - R_F (I_1 + I_C)$$

$$V_0 = - R_F \left(\frac{V_1}{R_1} + C_D \frac{dV_1}{dt} \right)$$

The output voltage therefore has two desired components, *i.e.* proportional to input and proportional to rate of change of input. The feedback resistor is necessary to give proportional addition and adjustment. If, as before, measured value and desired value inputs through resistors are applied then input voltage is deviation error signal and output voltage is signal to final control element. The phase advance network ahead of the amplifier gives attenuation across the CR circuit which requires compensation with increased gain at the amplifier. The R_2 desired value resistor is shown dotted. Derivative action is never applied on its own.

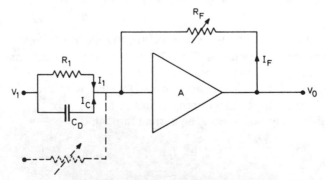

Fig. 11.4 ELECTRONIC PROPORTIONAL PLUS DERIVATIVE
CONTROL

ELECTRONIC COMPOUND CONTROLLER $(P + I + D)$

A controller action in which the output signal from the controller is the result of more than one operation on the deviation (error signal), in this case three, i.e. three term controller.

$$V_0 = - R_F \left(\frac{V_1}{R_1} + \frac{1}{C_F R_1} \int V_1\, dt + C_D \frac{dV_1}{dt} \right)$$

leading to:

$$V_0 = - \frac{R_F}{R_1} \left(V_1 + \frac{1}{C_F} \int V_1\, dt + R_1 C_D \frac{dV_1}{dt} \right)$$

The relation with the proportional, integral and derivative factors given previously and equations relating to Fig. 10.8 are obvious.

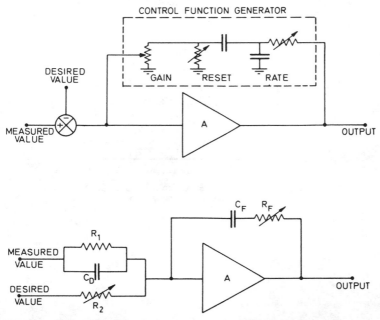

Fig. 11.5 COMPOUND $(P + I + D)$ ELECTRONIC CONTROLLER

Referring to Fig. 11.5:

The sketch shows the compound electronic controller, the action should be clear from previous diagrams and should be compared with Fig. 10.8. Similar remarks about interaction apply as for the pneumatic case. The upper sketch illustrates grouping to controller (note the summer, and the potentiometer gain adjustment) whilst the lower sketch is basic operational amplifier configuration.

Figure 11.6 is a typical response curve for $(P + I + D)$ action. Figure 11.7 is a simplified form of electrical circuit which can be compared with the previous sketch and to the complete electronic controller diagram given later in Chapter 12.

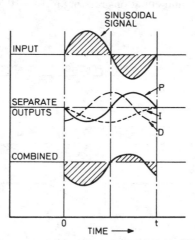

Fig. 11.6   RESPONSE TO SINUSOIDAL SIGNAL
(COMPOUND $P + I + D$)

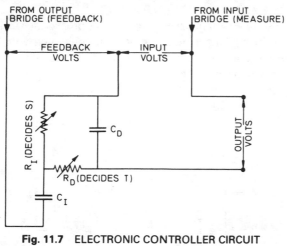

FROM OUTPUT
BRIDGE (FEEDBACK)

FROM INPUT
BRIDGE (MEASURE)

FEEDBACK VOLTS

INPUT VOLTS

C_D

R_I (DECIDES S)

OUTPUT VOLTS

R_D (DECIDES T)

C_I

Fig. 11.7   ELECTRONIC CONTROLLER CIRCUIT
(COMPOUND $P + I + D$)

Note:

With electric-electronic controllers:

1. For output voltage to be proportional to input voltage it is necessary to add a resistor (R_1) in the feedforward path to the amplifier (*i.e.* in series) and a resistor (R_F, adjustable) in the feedback path (*i.e.* in parallel).

2. If output volts are to be proportional to deviation, represented by error input volts (e), then error signal is applied as input to the circuit of 1.

3. For the addition of integral action to 2 above it is necessary to include a capacitor (C_F) in the feedback circuit in series with resistor R_F.

4. For the addition of derivative action to 2 above it is necessary to include a capacitor (C_D) in the feedforward circuit to the amplifier in parallel with the resistor R_1.

5. If $P + I + D$ action is required to combine 2, 3, 4, above, then

$$V_0 = - \left(\frac{R_F}{R_1} e + \frac{R_F}{R_1 C_F} \int e \, dt + R_F C_D \frac{de}{dt} \right)$$

BLACK BOX ANALYSIS

It is useful to summarise some of the preceding work in this chapter utilising this analysis approach, concerned with external relationships and not internal circuitry (which includes operational amplifier with negative feedback loop). Amplifier power supplies, and earthing, are important and actual voltage used depends on the amplifier and conditions of working.

Referring to Fig. 11.8 for illustrative connections:

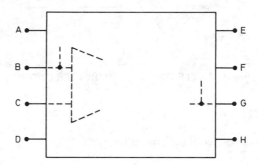

Fig. 11.8   BLACK BOX ANALYSIS (OPERATIONAL AMPLIFIER)

A is null offset   B amplifier inverting input (−)
C amplifier non-inverting
    input (+)   D supply (− 6V)
E spare terminal   F supply (+ 6V)
G output (feedback to B)   H null offset

Only two input signals (maximum), are considered to either B or C for simplicity of illustrations. Terminology for Fig. 11.8 should be clear except for null offset, which needs elaboration. In derivation of equations the input current to the operational amplifier is assumed negligibly small, and where resistances are specified equal that these are exactly so. This is practically not possible. To balance input resistances, and reduce such offset, terminal C (when not in use) is earthed through a resistance (R_E) of magnitude equal to the equivalent resistance of feedback and input circuits in parallel. To balance any remaining offset, and this is particularly important for integrating circuits, terminals A and H are used to give offset null. They connect a tapped (resistor) potentiometer across the ends of the amplifier to supply lead (usually negative) so giving compensation for inherent offset at amplifier input.

Consider now various configurations applied to Fig. 11.8:

1. Inverter $V_0 = -V_1$

Input voltage signal V_1 through a resistor R_1 to terminal B, terminal C earthed through a resistor R_E. $R_1 = R_F$. Output voltage signal V_0 at terminal G (also see Fig. 11.1, upper lead to A is inverting input, lower lead to A is non inverting input—shown earthed).

2. Inverter Summer $V_0 = -(V_1 + V_2)$

As for inverter but input signals V_1 and V_2 in parallel to B. $R_1 = R_2 = R_F$ (also see Fig. 11.2).

3. Scalar Inverter Summer $V_0 = -(aV_1 + bV_2)$

As for inverter summer but $a = R_F/R_1$, $b = R_F/R_2$, $R_1 \neq R_2$, R_F usually adjustable (see previous sketches).

4. Inverter Multiplier $V_0 = -at\,V_1$

As for inverter but input potentiometer (resistance R) connected across V_1 and earth, with a tapping lead to R_1 then to B. $a = R_F/R_1$, $R_1 \neq R_2 \neq R_F$, tapping ratio $t = r/R$, which is less than unity.

5. Inverter Divisor $V_0 = -V_1/t$

As for inverter but output potentiometer (resistance R) connected across V_0 and earth, with a tapping lead to make the feedback loop through R_F to B, tapping ratio $t = r/R$, which is less than unity.

6. Non Inverting Summer $V_0 = V_1 + V_2$

V_1 input through R_1 in parallel with V_2 input through R_2 into C, which is led to earth via resistor R_3, B earthed through resistor R_4. $R_1 = R_2$, $R_4 = R_F$, $R_3 = R_1/2$ (as two inputs).

7. Subtractor $V_0 = V_1 - V_2$

V_1 through R_1 to C, earthed via resistor R_3. V_2 through R_2 to B. All resistances equal.

8. Non Inverting Multiplier $V_0 = cV_1$

V_1 through R_1 to C. B earthed through R_4. $R_1 = R_4$. $c = (R_F + R_4)/R_4$.

9. Integrator

The circuit has already been sketched in Fig. 11.3. The amplifier C terminal could be earthed through R_E and terminals A and H connected.

The analysis could be further extended to two term and three term controllers.

TRANSISTOR STABILISATION CONTROL

The constant current (shunt) stabiliser utilising a zener diode has been described previously (Fig. 7.6). Voltage (or series) stabilisation based on an integrated circuit, *i.e.* transistor, gives better results. The principle is to use the transistor (T) as a variable resistor by a feedback control loop (see Fig. 11.9). Transistor resistance depends on emitter-base potential which is controlled by output voltage. Output (V_0) and standardised (zener) voltage (V) are compared and error voltage (e) is fed to the transistor base through a difference amplifier (A).

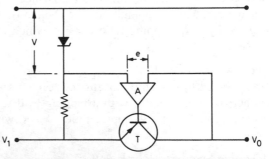

Fig. 11.9   TRANSISTOR STABILISATION CONTROL

INTEGRATED CIRCUITS

These were originally developed for non-linear or digital (logic) circuitry but are increasingly used for linear amplifiers, including operational amplifiers. The circuit components are produced by oxidising, etching and diffusion on a silicon chip (see Page 92) as shown in Fig. 11.10. A MOSFET (see Page 106) is often utilised within printed circuit boards. A microprocessor is a very large integrated chip circuit.

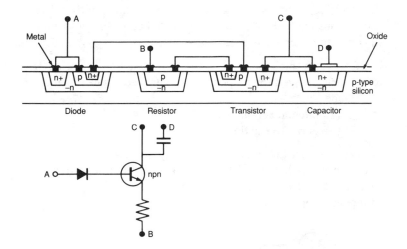

Fig. 11.10   INTEGRATED CIRCUIT COMPONENTS

TEST EXAMPLES 11

1. Describe a three term electronic controller. Show on a diagram the variation in controller output due to:
    (a) proportional action,
    (b) integral action,
    (c) derivative action,
    (d) combination $P + I + D$.

2. Describe, in detail, an electronic operational amplifier. Discuss the various methods of adjusting gain and comment on the uses of this device in control or measuring systems.

3. Show, by means of simple sketches, how an electronic controller arranged for proportional control action can be modified to include:
    (a) integral action,
    (b) derivative action.

4. Fig. 11.11 illustrates a time delay circuit which includes a rectifier and a rhyristor. The network varies lighting intensity i.e. it is a typical dimmer switch circuit. Describe in detail, the principle of operation.

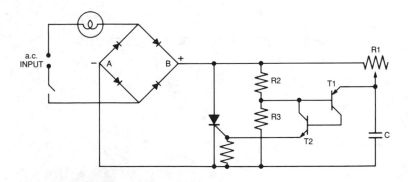

Fig. 11.11

CHAPTER 12

ACTUAL CONTROLLER TYPES

There are numerous forms of controllers produced by a large number of specialist instrument and control equipment manufacturers. As a representative selection of present well proven practice twelve types are given; one mechanical, one mechanical-hydraulic, one electro-pneumatic, three electronic and six pneumatic. In some controllers the action is given as $(P + I + D)$. It should be remembered that the most simple control suitable ought to be used and the controllers described are also available as simple (P) or $(P + I)$ or $(P + D)$. The description so given is mainly to cover the more involved case which can then be easily simplified.

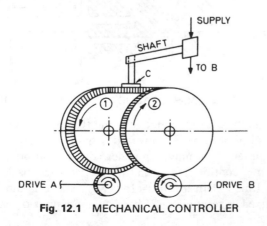

Fig. 12.1 MECHANICAL CONTROLLER

MECHANICAL CONTROLLER

Utilises a lever principle extended to a mechanical differential. A drives (1) anticlockwise and B drives (2) clockwise. If A and B have the same speed then C revolves but the shaft has no linear motion. If A tends to speed up then C follows (1) and the shaft moves linearly to alter the supply and increase the revolutions of B. A proportional action exists, synchronism reduces movement to zero, a distinct anti-hunt characteristic exists (see Fig. 12.1).

MECHANICAL-HYDRAULIC CONTROLLER (GOVERNOR)

The device is illustrated in Fig. 12.2.

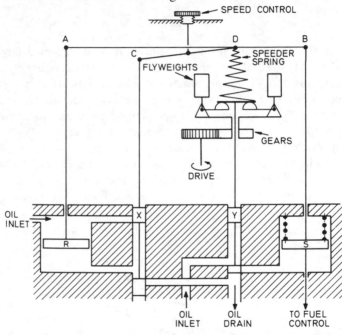

Fig. 12.2 MECHANICAL-HYDRAULIC CONTROLLER (GOVERNOR)

For say an increase in engine load the flyweights move radially inwards and pilot valve Y moves down so allowing the servo-power amplifier (S) to move up under admitted oil pressure. This increases fuel control setting and also rotates feedback link AB and reset link CD in an anti-clockwise sense about pivot A (initially fixed as reset piston R has equal pressures on each side).

Due to CD rotation pilot valve X moves down allowing R to move down due to oil escape to drain. As R moves down to a new equilibrium position, AB pivots about B and rotates CD clockwise so closing X, locking R and restoring D to its original position. The engine is now running at original speed but at a different load. The governor is $P + I$, i.e. isochronous. To reduce the governor to simple proportional pivot A is fixed, R and X eliminated, and link CD removed. The conical form spring gives linearity to the speed measuring system.

A similar device is shown in Fig. 14.9 and an electrical alternative in Fig. 14.10.

ELECTRO-PNEUMATIC CONTROLLER

The controller shown in Fig. 12.3 is a converter. A sliding contact resistor S is moved by the valve spindle to form the position feedback and the input measure signal is at contact M. If

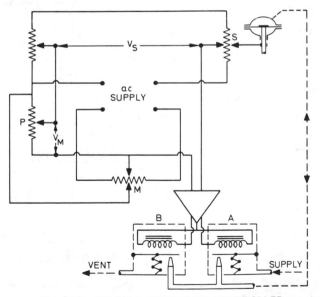

Fig. 12.3 ELECTRO-PNEUMATIC CONTROLLER

say $V_M + V_S > 0$ then relay A is energised and supply air flows to the diaphragm top to move the spindle down. If $V_M + V_S < 0$ relay B is energised and air from the diaphragm top is vented, so allowing the valve spindle to move up. Movement ceases as soon as V_M and V_S equate to zero when both nozzles are closed and no current flows. Proportional band, for the valve positioner, is adjustable at P.

ELECTRONIC CONTROLLER (1)

Referring to Fig. 12.4 (and also refer back to Fig. 11.7). Considering first just the proportional action. The deviation (by the transducer) causes movement of the adjustable rheostat at A so that A and B now have a voltage difference, *i.e.* the input Wheatstone bridge circuit is unbalanced, current therefore flows between A and B.

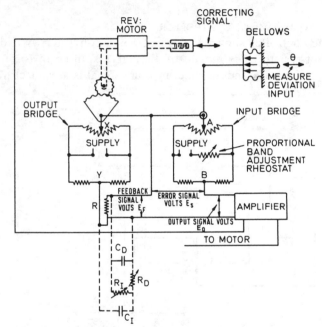

Fig. 12.4 ELECTRONIC $(P + I + D)$ CONTROLLER

This current produces an input voltage which is an indication of the error (deviation) magnitude from the desired value. Current flows to the amplifier, then to the electric motor. As the motor moves to alter the controlling element it moves the adjustable rheostat at X so that X and Y have a voltage difference, *i.e* the output bridge is unbalanced; current therefore flows between X and Y. This current is arranged to be opposite in direction to the current between A and B. The *fixed* resistance R therefore has opposing current flow and when the volt drop across the output bridge equals the volt drop across the input bridge (*i.e.* feedback equals input) the voltage across the amplifier is zero, motion stops. Therefore for each A position there is a *proportional* X position.

For output to equal zero, feedback volts equals supply volts and is opposite in direction, i.e. $E_F = -E_S$ or, as E_S is adjustable, due to the adjustable rheostat for proportional band, and introducing linkage and rheostat proportionality factors:

$$\Phi = -\mu\theta$$

To add integral action *replace R* by an adjustable resistor R_1 and a condenser of capacitance C_1. Thus there is a continuous adjustment of feedback. Current flow charges the capacitor which then resists further current flow.

To add derivative action *retain R* and *add* an adjustable resistor R_D and a condenser of capacitance C_D. The capacitor must be charged at a rate equal to the rate of change of input supply E_S.

To give $(P + I + D)$ *omit R* and *replace* by the dotted circuit containing C_D, R_D, C_1, R_1 as shown in the sketch. A more modern design of this type utilises photo cells and electronic valves (or transistors) but the principle is much the same as above.

Note:
Mathematics could be introduced to show:

$$E = E_F + E_C$$

$$E = E_F + \frac{1}{R_1 C_1} \int E_F \, dt \quad \text{(integral)}$$

and

$$E = E_F + E_D$$

$$E = E_F + R_D C_D \frac{dE_F}{dt} \quad \text{(derivative)}$$

E is voltage on output bridge.
E_C is volt drop across capacitor C_1.
E_D is volt drop across resistor R_D.

Possibly a more simplified sketch arrangement of an alternative type of electronic controller is shown in Fig. 12.5.

ELECTRONIC CONTROLLER (2)
Referring to Fig. 12.5.

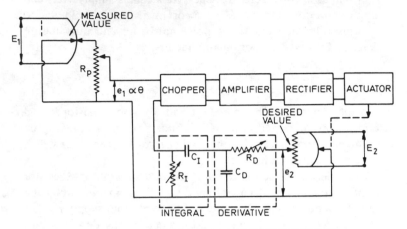

Fig. 12.5   ELECTRONIC $(P + I + D)$ CONTROLLER (2)

The detector element transmits a dc signal (measured value) which is usually amplified and may then be in the range 0–40 V, 0–10 mA. Similarly the desired value from a fixed resistor is supplied in opposition to the first signal. The two signals are compared and the difference between them is obtained by passing the two currents in opposite directions through a common resistor, thus voltage across the resistor is proportional to deviation. The resistor acts as a sensitivity control potentiometer in that the proportional band can be varied by varying the ratio between the two parts of the potentiometer, the variation is 2 to 200% range, or more. Thus pure proportional action is obtained with *direct* action from the measured value and *negative feedback* from the desired value.

The signal is then passed through a transistorised chopper (capacity modulator) to give ac which is then amplified in an ac amplifier and (if necessary) rectified for use at the actuator. This procedure is usually necessary as dc amplifiers are subject to drift which at zero true input would affect the deviation and hence the controller output.

The voltage developed across the correcting unit signal from the actuating unit is fed back through the derivative and integral networks in cascade. Some interaction occurs which is reduced by an auto-resistor in the derivative unit. Note that the final positioner action gives signal feed back whereas an alternative described in *all* controllers given previously has the feedback signal straight back after the amplifier itself. An obvious advantage of electronic controllers is their flexibility, high speed of signal transmission and high gain. As automatic controls develop more complex controller actions, further than $P + I + D$, will certainly be required and electronic controllers will provide the most convenient method. Voltage difference e_1 α θ and feedback voltage is e_2, equilibrium exists when $e_1 = e_2$. I and D actions are generated by resistance capacity systems C_1 R_1 and $C_D R_D$. Action times are adjusted by variable resistors R_1 and R_D.

With R_D equal to zero then $P + I$ action is generated only, *i.e.* integral action shunted via C_D. If R_1 is made infinite then $P + D$ action is generated, $C_D R_D$ and C_1 are in series.

ELECTRONIC CONTROLLER (3)

The circuit is much simplified to facilitate the description of this modern unit. Refer to Fig. 12.6:

The derivative (rate) circuit is shown at A, B, C. Time adjustment is by adjustable resistor-capacitor (A) to the measure input. Rate amplifier (B) is transistorised (solid state), two stage, *npn*, dc type, of overall gain about 10, and introduces a signal to the bridge network (D) proportional to the rate of change of measurement. Separate derivative addition gives no interaction and rapid action as the unit is in the forward loop and not in the feedback loop. A filter (C) gives smoothing (RC network).

The dc measure signal (10-50 mA) and set point signal are developed across resistors and any error signal will unbalance the bridge until the amplifier output and negative feedback rebalance. A resistor (K) decides the gain by controlling feedback so proportional action and variable bandwidth are achieved. The function of the bridge is to impress the resultant of error and feed-

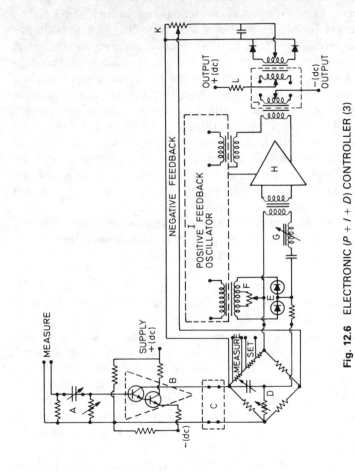

Fig. 12.6 ELECTRONIC ($P + I + D$) CONTROLLER (3)

back signals on the main amplifier (H) until equilibrium occurs.

The adjustable resistor and capacitor across the bridge determine the rate at which output changes (hence feedback) to drive the measurement to coincide with set point, *i.e.* integral (reset) action.

The controller amplifier part of the circuit is E, F, G, H, I, J. The main amplifier is similar to the rate amplifier but is four stage, ac, gain about 2000. The main amplifier bridge circuit consists of two diodes (E) arranged so that their capacitance change is proportional to dc voltage change across them, initial set bias and arranged in opposition for large unbalance to small signal, plus the split inductive transformer winding (F). Error input causes amplifier bridge unbalance (due to capacitance change) and an ac mV signal will enter H due to the oscillator (I). H is providing a positive feedback oscillating circuit for bridge excitation developed in the oscillator loop, tuned resonant circuit for bridge (G).

Output level is demodulated and raised by a two stage output amplifier (J) to an output signal in the range 10 to 50 mA, dc power supply, external load (L) and negative feedback (2-15 V) via diodes and resistor K. Note transformer couplings to isolate controller input and output circuits. External switching is provided with internal circuitry for set point change, limits automanual transfer, etc.. Essentially the unit described is a chopper-type dc amplifier fully transistorised. It consists of a transistor input chopper, a high gain ac amplifier, and a transistor output chopper (demodulator) with feedback.

PNEUMATIC CONTROLLER (1)

This unit gives (P), $(P + D)$, $(P + I)$, $(P + I + D)$ control actions and also provides addition (or subtraction), multiplication (or division) and averaging computing actions, as may be required. *This controller is an ideal example to illustrate all the basic actions in as simple a manner as possible.*

Referring to Fig. 12.7.

The principle is that of force-balance using the simple lever principle. Four bellows act on the beam (lever) and variations of

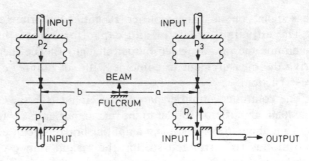

PRINCIPLE

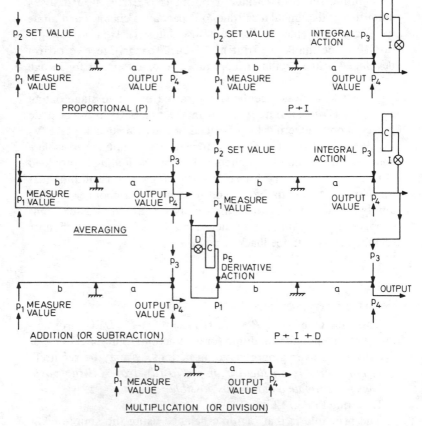

Fig. 12.7   PNEUMATIC CONTROLLER (1)

bellows forces or level fulcrum ratio ($a : b$) will affect the magnitude of the output signal.

$$(p_4 - p_3)a = (p_1 - p_2)b$$

$$\therefore p_4 = \frac{b}{a}(p_1 - p_2) + p_3$$

variations of a, b, p_1, p_2 or p_3 obviously affect the value of p_4 (output).

When used for proportional action:

$$p_4 = \frac{b}{a}(p_1 - p_2) \quad \text{with } p_3 = 0$$

i.e. difference between set value and measure value yields a proportional output signal which is adjustable by the $a : b$ ratio.

When integral action is added:
the restrictor I and capacity tank C give the necessary integral action via p_3 bellows.

When derivative action is also added:
the restrictor D and capacity tank C give the necessary derivative action via an extra p_5 bellows on the extra (lower) totaliser. The $P + I$ output from the upper totaliser p_4 bellows is fed to the p_3 bellows of the lower totaliser to give $P + I + D$ output from the lower totaliser p bellows.

For averaging:

Taking $a = b$ then as $p_2 = p_4$
$$p_4 = \tfrac{1}{2}(p_1 + p_3)$$

For addition:

Taking $a = b$ then as $p_2 = 0$
$$p_4 = p_1 = p_3$$

Subtraction can be arranged utilising the other bellows p_2 in place of bellows p_3.

For multiplication:
with $p_3 = p_2 = 0$
$$p_1 b = p_4 a$$

$$p_4 = \frac{b}{a} p_1$$

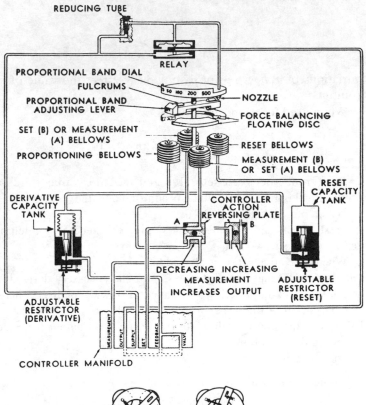

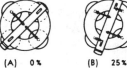

(A) 0 %   (B) 25 %

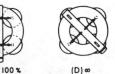

(C) 100 %   (D) ∞

Fig. 12.8   PNEUMATIC CONTROLLER (2)

and ratio ($a : b$) decides multiplication factor if over unity. Division can be arranged by making the ratio ($a: b$) less than unity.

PNEUMATIC CONTROLLER (2)

This unit gives (P + I + D) actions (Fig. 12.8).

Forces exist due to four bellows on the force balancing floating disc which acts as the flapper, the resultant of moments of bellows forces determines the throttle position. With the fulcrums over the proportional and reset bellows there is no feedback effect and the distance between the centre line of the adjusting lever and the other two bellows is a maximum so giving zero percent proportional band (see A). With the centre line of the adjusting lever about 1 unit from the proportional bellows and about 4 units from the set bellows gives a 25% proportional band (see B). 100% proportional band exists for C. Infinite proportional band exists for D. Note the reversed controller action available if required. Derivative addition gives delayed feedback with differential across the resistor where flow is proportional to rate of change of deviation. Integral addition gives the usual delayed feedback, on the reset bellows.

PNEUMATIC CONTROLLER (3)

Referring to Fig. 12.9.

The sketch is for a proportional controller but other type units are available with integral or derivative action added. The sensing element, in this case thermo-sensitive system, on a change of conditions will, via the Bourdon tube, alter the flapper position. This decides the control pressure, this pressure is led back via the proportioning orifice and sensitivity adjustment to the proportioning bellows. For maximum sensitivity this pressure feedback is vented and no pressure acts on the bellows. If sensitivity is decreased the resulting pressure build up on the proportional bellows contracts the bellows and moves the nozzle away from the flapper. This means a greater flapper travel is required to close the nozzle, *i.e.* wider proportional band. The relay pilot valve assembly shown has a null position non-bleed action. When the flapper approaches the nozzle the control pressure increases, this lifts the primary diaphragm causing the valve to open and allow air pressure above the secondary diaphragm to the outlets (3). Increase of pressure above the sec-

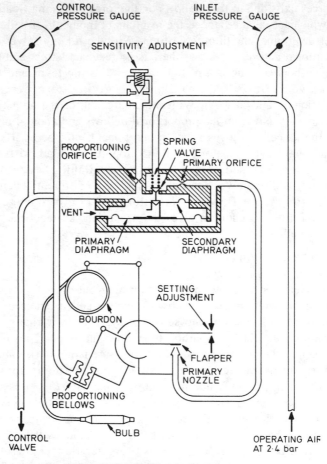

Fig. 12.9 PNEUMATIC CONTROLLER (3)

dary diaphragm will tend to balance pressure under the primary diaphragm, at balance the diaphragms return to the original position, the valve closes on both seats, and balanced pressure acts on the control valve. For flapper travel away from the nozzle then pressure above the secondary diaphragm will vent until balance is restored.

PNEUMATIC CONTROLLER (4)

The relay (Fig. 12.10) receives a proportional signal as input and modifies it to proportional plus integral. The relay consists of four chambers isolated from each other internally by metallic diaphragms connected to a central post, spring loaded at one end and operating inlet and exhaust valves through a beam at the opposite end. Alteration in chamber A pressure will affect a similar

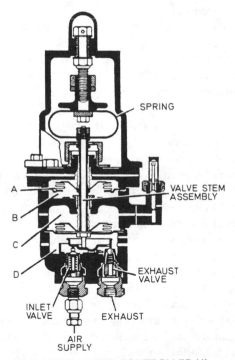

Fig. 12.10   PNEUMATIC CONTROLLER (4)

change in chamber D due to the repositioning of the valves caused by beam movement from the bellows. Flow also occurs to C via the restrictor throttle valve, as C is at a lower pressure, so giving a similar effect to the initial down movement but at a rate dependent on the deviation, *i.e.* integral action (adjustable by throttling valve). This regenerative effect continues until there is a return to the required setting when the supply proportional controller restores A chamber to balance giving a balance of C and D chambers and repositioning at the new required position. The spring maintains a given set value, at equilibrium both valves are slightly open.

The relay is easily modified to an averaging relay, *i.e.* taking two signals, combining and giving a resultant output. Loading pressure comes into A as before and also in through the throttling valve to C directly. Thus two effects are combined, the C signal can employ any time delay dependent on the valve orifice opening. C and D are not connected.

Such a relay controller is characteristic of many similar designs usually called **stack** type (often without use of beams, *i.e.* valve and diaphragm action in one line only). They are sometimes termed **blind** controllers as they are often sited close to the control function to minimise distance-velocity lags.

PNEUMATIC CONTROLLER (5) (FUEL-AIR RATIO)

Change in combustion air flow is measured in terms of pressure difference across the air register, and is transmitted via the large bellows to the ratio beam (Fig. 12.11). Change in fuel oil pressure, caused by the master pressure controller due to variations of steam pressure, is fed to the smaller OF bellows. These two signals are in opposition when applied to the beam system.

Between the beams there is a movable roller fulcrum the movement of which, by the ratio adjustment screw, gives different equilibrium conditions and the ratio is indicated on the ratio scale. Beam lever position operates a linkage to the pilot valve which varies control air output signal. This output signal is fed to the averaging relay where it "trims" the signal being fed through to the air damper actuators. The adjustable proportional band and negative feedback bellows should be noted. This type of controller utilises proportional control only.

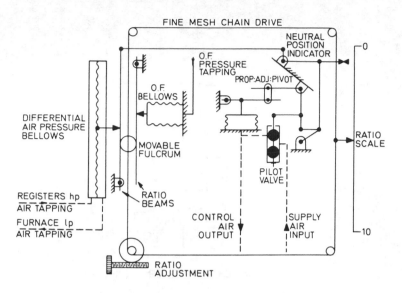

Fig. 12.11 PNEUMATIC CONTROLLER (5) (FUEL-AIR RATIO)

The correct fuel-air ratio can be maintained irrespective of the numbers of burners in use provided that air registers are closed on burners which are not in use.

PNEUMATIC CONTROLLER (6) (VISCOSITY)

The viscosity sensor has been described previously (Fig. 5.2). The high pressure connection (+) and low pressure connection (−) is led to a dp cell. Consider now Fig. 12.12:

Differential pressure is applied across the diaphragm D of the transmitter (cell). Increasing differential pressure (increasing viscosity) causes the diaphragm and balance beam to move to the left. The inlet supply nozzle B is opened by the flapper F which allows build up of air pressure in the feedback bellows B. This gives a restoring action on the balance beam until equilibrium is again reached. Discharge nozzle A is shut. Air pressure in the feedback bellows is output signal of the controller through C to a diaphragm valve regulating steam to the oil fuel heater.

For decreasing viscosity, discharge nozzle A is opened giving air bleed, and inlet nozzle B is closed. At equilibrium nozzles are

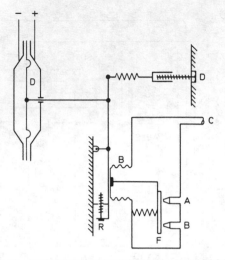

Fig. 12.12 PNEUMATIC CONTROLLER (6) (VISCOSITY)

virtually closed which reduces air wastage. Zero adjustment is at Z and range adjustment at R. The proportional action is readily extended to include integral action by adjustable reset control on the feedback bellows. Sensor and controller are described as a control circuit, see Fig. 13.19.

TEST EXAMPLES 12

1. Sketch and describe an instrument to maintain the viscosity of a fuel at a constant value. Explain how it corrects any deviation of the viscosity from the desired value.
2. Explain why load sensing governors are usually fitted to engines driving alternators. Sketch a governor for this duty and explain its action.
3. Describe, with the aid of a sketch, a specific type of three term pneumatic controller. Discuss how the separate control actions are generated and adjusted.
4. An electronic controller incorporates an integrating and differentiating network. Sketch each circuit. A square wave input is applied first to one circuit and then to the other with display on a CRO. Illustrate the input and output (2) wave forms on a common time base diagram.

CHAPTER 13

TYPICAL CONTROL CIRCUITS

The number of different control loops utilised is large. Applications for each of steam, motor (IC engine), general, engineering knowledge sections will now be considered. Much process control is still pneumatic, illustrations are mainly biased to this type. Electronic sensing and control devices can easily be substituted but the final power control element is often pneumatic. Control of displacement, velocity and acceleration using electrical-electronic servo-mechanisms are detailed in the next chapter.

STEAM PLANT

Auto-combustion and feed system control have been used for many years. Modern sophisticated sub-systems for temperature, pressure, level, flow, etc., are interlinked into boiler and turbine overall control systems suitable for remote and transient conditions.

(1) EXHAUST RANGE PRESSURE CONTROL

This utilises sequence operation with valve positioners, see Fig. 13.1.

Range pressure (1 bar) is sensed and converted to a pneumatic pressure signal by the transducer. This signal is transmitted and compared with the set value and any deviation causes the controller to give an output signal change proportional to the deviation. Output signal 2 to 1·75 bar gives dumping to condenser, 1·75 to 1·5 bar 1p bleed is fed in, 1·5 to 1·25 bar desuperheated steam is fed in

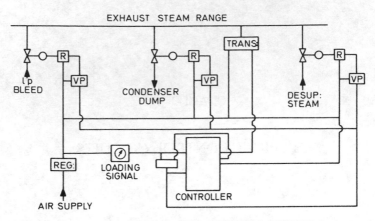

Fig. 13.1 EXHAUST RANGE PRESSURE CONTROL

whilst bleed remains full open. Thus exhaust range is constant pressure maintained. On the diagram R is for relay and VP is for valve positioner. This is split range control.

(2) TURBINE GLAND STEAM PRESSURE CONTROL

Such an arrangement is given in Fig. 13.2 from which it will be noted that gland steam pressure is sensed and supply either increased or dumped.

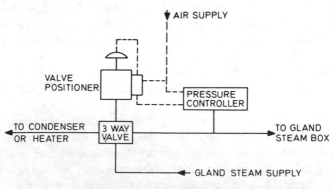

Fig. 13.2 TURBINE GLAND STEAM PRESSURE CONTROL

(3) SOOTBLOWER CONTROL

The sootblower system utilises air for both control and blowing. Operational rotation of the blower head is achieved by means of an air piston ratchet mechanism. The sequence of operation is governed by the distributor operated by a similar air piston ratchet mechanism. The supply air to both pistons comes from the pilot

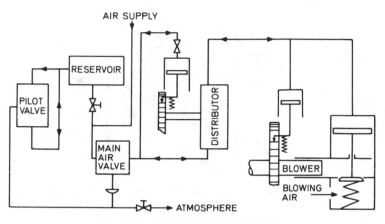

Fig. 13.3 SOOTBLOWER CONTROL

valve whose operation is dependent on the charging and discharging rate of the air reservoir. Adjustable control orifices are provided at entry to reservoir (charging) and on the atmosphere line (discharging). Each impulse or air puff blast rotates both ratchet gears by one tooth and gives a blowing blast for a few seconds (Fig. 13.3).

(4) CONDENSER CIRCULATING WATER TEMPERATURE CONTROL

This system utilises constant pump speed with water recirculation. For reduced power or low sea temperature operation the condenser may be operating far from design conditions. Excessive high vacuum results in possible turbine erosion, and low temperatures give excessive undercooling of condensate which reduces plant efficiency.

Referring to Fig. 13.4:

A fall of sea temperature would generally be arranged to decrease the air loading pressure, giving bellows expansion, which

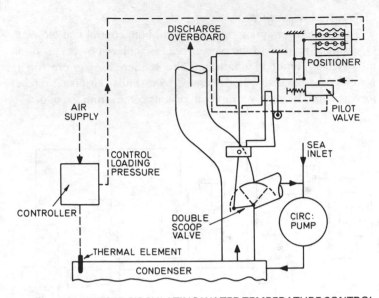

Fig. 13.4 CONDENSER CIRCULATING WATER TEMPERATURE CONTROL

through the positioner allows air to the top of the servo piston so allowing more of the scoop into the discharge pipe and giving more recirculation.

(5) STEAM TEMPERATURE CONTROL

Referring to Fig. 13.5:

Superheat control is based here on the amount of steam flow through the attemperator. The sensing element input signal to the steam temperature transmitter (T) is directed to the recorder-controller (C), which is often three term. Output signal from this controller is combined in an adding relay (A) with the output signal from a steam flow transmitter (F). Relay output signal passes through the control station (S) (hand-auto) to operate the valve positioner with linked control valves to vary attemperator flow rate.

Two element control allows more effective operation during transients. Increased steam flow would reduce temperature without the second element action which would be reducing flow through the attemperator. Split range control (two valves and positioners)

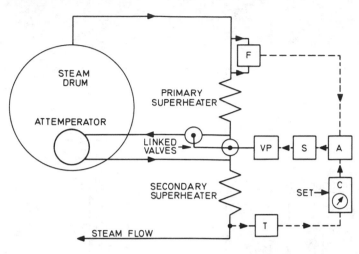

Fig. 13.5   STEAM TEMPERATURE CONTROL

could be used in place of two linked valves from one positioner.

Note.

Three *term* usually means $P + I + D$ and three *element* usually means three variable combination in a controller *e.g.* pressure, level, flow.

(6) STEAM FLOW/AIR FLOW RATE CONTROL

Refer to Fig. 13.6:

Steam flow rate is sensed at the orifice plate with tappings to a steam flow transducer consisting of a dp cell (1) usually supplied with condensed water. The cell would incorporate a square root eliminator, perhaps mercury well type with mechanical linkage to variable inductance (2) operated amplifier (3). The air flow rate is similarly sensed. Outputs from both dp cells are fed to an electronic computing relay whose output signal is related to required air flow for the measured steam demand. A tapping can be arranged to the burner fuel supply controller. The computing relay output signal enters a three term electronic controller. The electrical signal output from the controller operates the final control element. This element

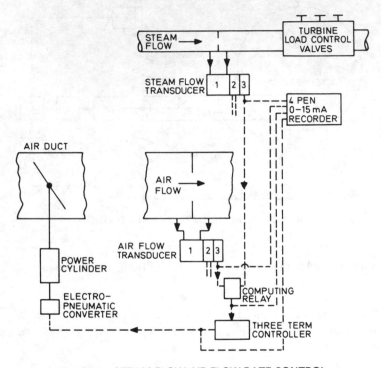

Fig. 13.6   STEAM FLOW/AIR FLOW RATE CONTROL

is a damper in the inlet air duct. An electro-pneumatic converter and power cylinder are required. Note direct signal measures of the four variables at pen recorder.

(7) BRIDGE CONTROL (TURBINE MACHINERY)
Instrumentation and Alarms.

For the bridge console the least instrumentation and alarm indicators the better, alarms should be essentials only and instruments only those vitally necessary.

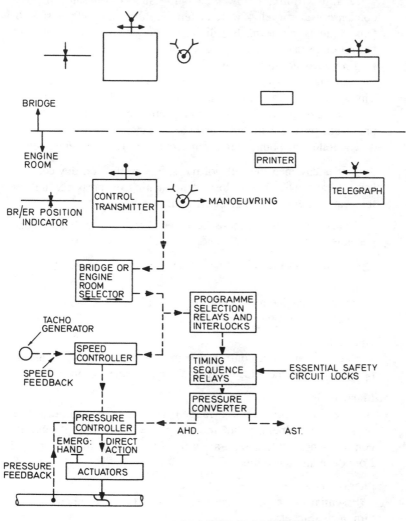

Fig. 13.7 BRIDGE CONTROL (TURBINES)

Suggested alarms could be:

1. High salinity, 2. low feed pump suction pressure, 3. high condenser water level, 4. low vacuum, 5. lubricating oil pressure, 6. tank contents low level. For direct instruments opinion is divided but no more than say another six indications should be necessary. Engine console and alarms would obviously provide full in-instrumentation. A typical simplified system is given in Fig. 13.7. This system has direct control at the steam manoeuvring valves.

This is essentially a combination electro-pneumatic although all-pneumatic or all-electric can easily be arranged.

The following points with reference to Fig. 13.7 should be noted:

1. It is assumed that all normal safety protection devices and control are provided, *e..g.* loss of lubricating oil pressure, high or low water level in boilers, electrical failure, etc.

2. Subsidiary control loops have been omitted, *e.g.* evaporators, generator, etc.

Consider now the individual aspects relating to Fig. 13.7:

Selector
Bridge or engine room control can be arranged at the selector in the engine room. When one is selected the other is ineffective.

Duplication
Both transmission control systems are normally identical and operation of the one selected gives slave movement of the other.

Manoeuvring
A separate sequence is arranged, for example opening astern master valve, opening turbine drains, etc., but if this separate control is not applied separately it will be automatically applied from the main transmitter.

Interlocks
Essential blocks are necessary such as no valve opening with turning gear in, etc.

Programme Relays
The correct sequence operation is arranged for the various actions for example, manoeuvring conditions satisfied before moving on to say valve opening sequences, etc.

Timing Relays

To prevent excessive speed changes, by too rapid signals, which would endanger engines and boilers. For example half speed in say ½ minute, emergency swings from say full ahead to full astern allow astern braking steam usage, rotation slightly ahead and astern at say 3 minute intervals of a sustained stop of engines, etc. (manoeuvring valves now usually cam operated sequential).

Essential Safety Locks

Override on timing by such elements as low boiler water drum level, low turbine inlet steam pressure, etc.

Emergency Control

Direct hand control of manoeuvring valves camshaft.

Local Control

Independent power control at the actuators themselves.

Feedbacks

The steam pressure feedback gives accurate positioning for pressure of the actuator. Speed feedback is arranged so that a difference of speed between measured and desired values causes an additional trimming signal to the controller. This may be necessary as pressure and speed are not well correlated at low speeds.

Outline Description

The following is a brief description of one type of electronic bridge control for a large single screw turbine vessel to illustrate the main essentials. Movement of a control lever modifies the output of an attached transmitter (electronic signal 0-10 mA dc). The transmitted signal is passed, via override, alarm and cut out units, to the desired flow module which is connected to a time relay and feeds to the controller. The electronic controller compares desired speed with actual speed as detected by a tachometer generator and dc amplifier. The correct controller signal is passed to the manoeuvring valve positioner from which a return signal of camshaft position is fed back to the dc amplifier, thus giving the command signal to the actuator reversing starter. The two control levers are independent and do not follow each other, an engine room override of bridge control is supplied. The rate of valve opening is controlled by the actuators so that too rapid valve opening is prevented by a time delay, full normal valve operation shut to open, or vice versa, occurs in about 1 minute, this can be reduced to about 20 seconds in emergency by full movement of the telegraph from full speed direct to or through stop. A near linear rev/min to control lever position exists. Auto-blast refers to the automatic time delay opening of the ahead manoeuvring valve for a short period after a certain length of time stopped - this has an override cut out for close docking, etc.

IC ENGINE PLANT

IC engine plant now utilises general (mainly thermal) auto-control to a great extent and manoeuvring systems, etc., have always involved fairly detailed devices. The efficiency is virtually inherent in the design, however auto-control can still give improvements to efficiency of operation.

(1) JACKET (OR PISTON) TEMPERATURE CONTROL - SINGLE ELEMENT
    Referring to Fig. 13.8:

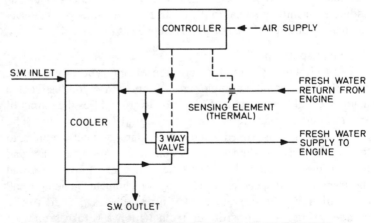

Fig. 13.8 JACKET (OR PISTON) TEMPERATURE CONTROL –
SINGLE ELEMENT

The sensing element may be on the supply or return line to the engine, there are certain advantages for each case. The 3 way valve (2 entry and 1 exit) varies the re-circulation or supply to the cooler and functions as a mixing valve. This valve may be of the rotary cylinder type or diaphragm operated type. Integral action is usually incorporated in the controller as offset may be appreciable otherwise (up to 9°C). This system is often preferred. An alternative arrangement is to throttle the sea water supply for the cooler. This gives a big capacity lag in the system as one variable (sea water) controls the other variable (fresh water). Valve selection is most important. Maximum pressure and temperature, maximum flow rate, minimum flow rate, valve and line pressure drops, etc., must be carefully assessed so that valve gives the best results. The controller shown in Fig. 13.8 is pneumatic but an electronic controller with a rotary valve is also common. Engine coolant is shown as fresh water but could equally well be oil. Most loops on a motor ship are single element, for example temperature control of lubricating oil, fuel, air, water, etc.

(2) TWO ELEMENT COOLING LOOP

More sophisticated loops are sometimes found on auxiliary boiler controls (pressure, level, etc.) as well as main engine coolant.

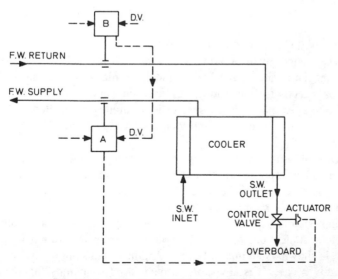

Fig. 13.9 TWO-ELEMENT COOLING LOOP

In the latter case a two element type may be advantageous as heat transfer rates are high and sea water temperature becomes more critical than in smaller single element loops. The two variables involved are engine load and sea water temperature.

If the engine is considered to be at a fixed load then by reference to Fig. 13.9 it is seen that the water inlet temperature is fixed by the set value of the controller A which accounts for water temperature changes (controller A is "slave").

Now if the engine load changes the inlet water temperature should change, *i.e.* the lower the load the higher the water temperature. This is achieved by changing the desired value of controller A according to the engine load variations. Controller B (the "master") provides an indication of engine load by measuring the return water temperature. Controller B signal changes the set value of controller A (cascade control). A fresh water heater may be placed in the engine supply line with heat input controlled by controller A. Split range (level) control allows heat input at low coolant temperatures and cold input at high coolant temperatures. The system works equally well for oil coolant. The actuator can be operated by local or remote control with the controllers out of operation.

(3) OVERALL COOLANT SYSTEM CONTROL

The sketch given previously in Fig. 13.8 for jacket (or piston) coolant temperature control is typical. Similar systems can apply almost exactly for lubricating oil, turbo blowers, scavenge air, fuel valve, etc., coolants. Each system is separately controlled so that adjustment can be made to individual systems without affecting the others.

One obvious arrangement for sea water supply is to utilise independent tappings (both main engines and auxiliaries) from a continuously circulated **ring main** - in this design individual sizes and flow rates require very careful investigation. A typical ring main system is given in Fig. 13.10. For simplicity no duplication of coolers, indication of individual valves, strum or strainer boxes, etc., is shown. Note that only main engine circuits are controlled in this given case. It is advisable to have say four pumps capable of operation on the main, the pumps preferably divided with say two

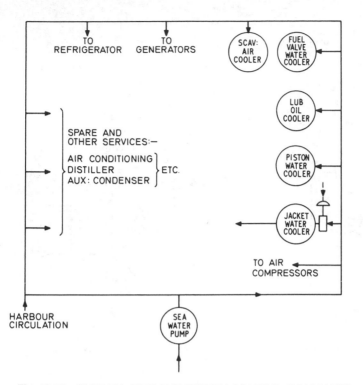

Fig. 13.10   OVERALL COOLANT SYSTEM CONTROL (RING MAIN)

constant speed and two controllable speed to allow flexibilty, harbour circulation is best arranged by a feed into the main from say the ballast pump. The individual control is shown on the jacket cooler only for illustration. The control here is direct control of the sea water quantity utilising two-way diaphragm control valves, all main engine circuits shown in the diagram are so controlled, any auxiliary circuits can be similarly controlled if required. Certain discharges, where convenient, can be combined to reduce the number of shipside valves.

Another alternative is a series or parallel arrangement

The two diagrams given in Fig. 13.11 illustrate series circulation with three way valves with bypass, and parallel circulation with two way valves with direct supply, the latter arrangement requires a

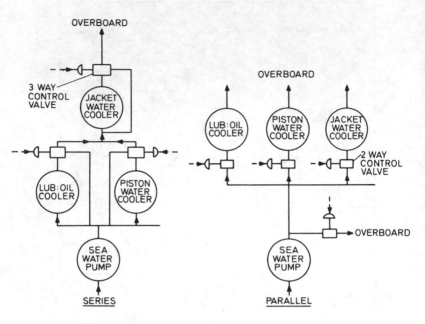

Fig. 13.11   OVERALL COOLANT SYSTEM CONTROL
(SERIES & PARALLEL)

satisfactory system pressure control such as control flow from pump or dump of excess water (as shown). Grouping of the coolers and choice or combination of systems can be arranged dependent on flow and temperature considerations. No duplication, etc., is shown.

(4) BOILER OIL PURIFICATION CONTROL SYSTEM FOR IC ENGINE

This system shown in Fig. 13.12 is designed to maintain a working level in the boiler oil service tank to the main engine. The oil supply from the dirty oil tanks continuously passes through a self cleaning purifier. The oil fuel heater (at purifier, or if also provided on the main engine supply rundown) can easily be arranged to give fixed oil temperature or controlled viscosity in a similar manner to the previously described temperature flow control system.

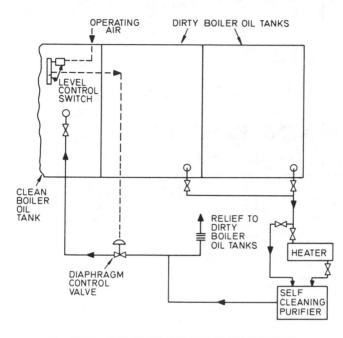

Fig. 13.12   BOILER OIL PURIFICATION CONTROL
SYSTEM FOR IC ENGINE

(5) WASTE HEAT FLASH EVAPORATOR CONTROL

This system is shown in Fig. 13.13. Waste, or low grade heat, from engine coolant has a good energy potential. Fresh water jacket (or piston) coolant evaporates sea water in the second stage heat exchanger which is condensed at about 0·1 bar in the first stage pre-heater and removed by the distillate pump.

When the pressure sensor-transducer (P) allows the controller (C) to operate the sea and coolant inlet valves, vapour production starts. Control is by measurement of made water flow at the flow sensor transducer (F) and the signal allows the controller to regulate water inlet valves accordingly.

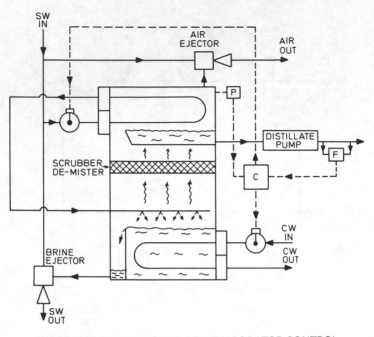

Fig. 13.13   WASTE HEAT FLASH EVAPORATOR CONTROL

(6) BRIDGE CONTROL (DIRECT REVERSING IC ENGINE)

It is suggested that the section on bridge control for turbines should be read again as there are many obvious similarities.

Instrumentation and Alarms

Minimum usage, suggested alarms (*bridge console*):

1. Low starting air pressure, 2. lubricating oil discharge pressure and temperature, 3. cooling water discharge pressure and temperature, 4. tank contents level gauge, 5. fuel oil discharge pressure and temperature, 6. scavenge belt pressure.

A further six instruments could be provided.

Note:

All normal protective devices are assumed, subsidiary control loops are not considered.

Refer to Fig. 13.14 and consider the following:

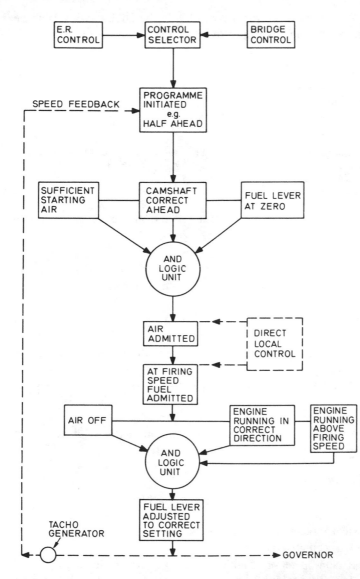

Fig. 13.14   BRIDGE CONTROL (DIRECT REVERSING IC ENGINE)

Selector
Bridge or engine room control is in the engine room. With one selected the other is inoperative.

Duplication
Both transmission control systems are identical master and slave functions as selected.

Interlocks
Essential blocks are necessary, such as no action with turning gear in, etc.

Programme and Timing Relays
Consider say a requirement of "half ahead". The programme must satisfy a sequence such as:
1. Fuel admission check for zero.
2. Air lever to ahead.
3. Sufficient air and camshaft direction checks.
4. Air admitted.
5. Adjustable delay period for firing speed.
6. Fuel admitted.
7. Delay checks for air off, correct direction, rotation above firing speed.
8. Adjust fuel for set value speed.

Essential Safety Locks
Override on timing by such elements as low lubricating oil pressure, low cooling water pressure, etc.

Emergency and Local Control
Directly at the engine controls themselves.

Further Protective Considerations
1. Speed governor.
2. Non-operation of air lever during direction alteration.
3. Failure to fire requires alarm indication and sequence repeat with a maximum of say four consecutive attempts before overall lock.

4. Movement of control lever for fuel for a speed out of a critical speed range if the bridge speed selection is within this range.

5. Emergency full ahead to full astern, etc., actions, must have time delays to allow fall of speed before firing revolutions, astern air open, engine stop, correct astern timing and setting.

Outline Description

The following is a brief description of one type of electronic-pneumatic bridge control for a given large single screw direct coupled IC engine to illustrate the main essentials. The IC engine lends itself to remote control more easily than turbine machinery.

Movement of the telegraph lever actuates a variable transformer so giving signals to the engine room electronic controller which transmits, in the correct sequence, a signal series to operate solenoid valves at the engine. One set of solenoid valves controls starting air to the engine while a second set regulates fuel supply, the latter via the manual fuel admission lever, is coupled to a pneumatic cylinder whose speed of travel is governed by an integral hydraulic cylinder in which rate of oil displacement is governed by flow regulators. This cylinder also actuates a variable transformer giving a reset signal when fuel lever position matches telegraph setting.

With the engine on bridge control the engine control box starting air lever is ineffective and the fuel control rack is held clear of the box fuel lever. Normal fuel pump control is eliminated and fuel pressure, in a common rail, is automatically adjusted to speed and load by a spring loaded relief valve. Engine override of bridge control is provided.

The function of the electronic controller is to give the following sequence for say start to half ahead: Ensure fuel at zero, admit starting air in correct direction, check direction, time delay to allow engine to reach firing speed, admit fuel, time delay to cut off air, time delay and check revolutions, adjust revolutions. Similar functions apply for astern or movements from ahead to astern directly. Lever travel time to full can be varied from stop to full between adjustable time limits of $\frac{1}{2}$ minute and 6 minutes. Fault and alarm circuits and protection are built into the system.

(7) REMOTE CONTROL (DETAIL - IC ENGINE)

Pneumatic reversing and starting systems have been an integral part of IC engine equipment for many years. Figure 13.14 and description are fairly general and it is now appropriate to extend this to a more detailed arrangement.

Consider Fig. 13.15 and the following nomenclature:

1. telegraph receiver; 2. speed lever; 3. start lever; 4. shut off lever; 5. bridge maximum speed lever; 6., 7., 8., 9., 10., 11. selector pilot valves; bridge or control room, ahead, astern, speed, start, maximum speed; 12. telegraph transmitter; 13. speed cam; 14. programme motor drive; 15. speed cam (fine control); 16. control cam ahead-astern; 17., 18., 19. selector pilot valves; speed, astern, ahead; 20., 21., 22. control valves; flow, speed, flow; 24. solenoid valve; 25. timing volume; 23., 26., 27. double check valves. There are ten relay valves, two each ahead and astern plus, U, V, W, X, Y, Z. A is air supply, B outlet lifts engine speed lever handgrip out of gear to allow remote operation, C outlet to speed set, D from and to direction interlocks, E outlet to reversing interlock, F outlet to starting servo, G outlet to reversing servo, L to solenoid valve.

With the engine telegraph set at remote control position the control air at a pressure of about eight bar flows through line A to all nine pilot valves (as shown full lines) if lever 4 is opened. There is an additional section on the engine telegraph for remote control and also on the control room telegraph for bridge control position. As 4 is open air passes through B, C, D, E to operate interlocks. Air also passes to the upper ahead or astern direction selector relay valves, partly clears speed set relay valve X for operation (via 21), and clears the starting routine at X and Z via 24 (from line L).

Operation of 1 selects either control room or bridge and output from 6 loads either U or V accordingly. Ahead lines 7, 19 lead to V and one as appropriate feeds top ahead relay valve and goes out via line G to position the reversing servo at the engine (astern similarly through 8, 19 via U and top astern relay valve to other line G).

Speed set from 2, 9 leads to W (directly loaded from control room) (via 6 or from bridge via 17, 5, 11, 20) thence to X and line C to act on speed set servo.

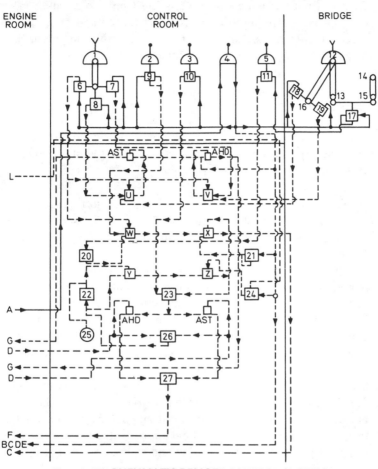

ENGINE ROOM   CONTROL ROOM   BRIDGE

Fig. 13.15   PNEUMATIC REMOTE CONTROL SYSTEM

Start from 3, 10 through 23 and either of lower ahead-astern relay valves past 27 via line F to engine starting cylinder servo (depending on loading signal from either line D direction interlocks). 23 is also cleared from 26, through Y and Z, with built in flow and timing control at 22 and 25.

The engine will for example reverse, start and gradually reach selected speed in an automatically programmed sequence. Two fine set speed buttons allow speed variation from fixed values. The system has built in avoidance of critical speed ranges and acceleration to full speed when on bridge control. This pneumatic system is readily adaptable to electronic signals to servos.

GENERAL PLANT

General control applications have always been used and the ship steering gear control system is perhaps a classic example. Early control components included the safety valve and Watt governor. A selection of typical systems is now given.

(1) AUTO-COMBUSTION AND ATTEMPERATOR CONTROL SYSTEM

For simplicity no feed water control or similar controls are shown in this system but they are commonly fitted in practice (Fig. 13.16).

For main propulsion steam turbine drive wide range fuel oil burners are a very desirable feature with automatic combustion control as they allow manoeuvring without the need to change burners. Burners must have a high turn down ratio. Mechanical atomisation burners have a turn down ratio of about 2:1 whereas steam atomisation assisted burners have a turn down ratio of about 20:1. The "steam" burners however induce a water loss of about 0·75 per cent of steam flow. The ideal arrangement is to utilise "steam assisted" burners during manoeuvring and "mechanical" burners at full power. This gives full automatic control over the manoeuvring range. An automatic dumping valve to the condenser ensures a nominal steam flow when the main turbines are stopped. Until recently automatic combustion control was only used at full power, in fact the main need and modern application is when load changes, *i.e.* manoeuvring transients.

For IC engine plant especially in view of the Clean Air Act, combustion control of boilers is virtually essential.

Fig. 13.16 is fairly simple in principle and can be considered suitable for auxiliary boiler practice in motorships. The detail should be taken in conjunction with the following system on the more fully automated auxiliary automatic boiler.

Referring to Fig. 13.16.

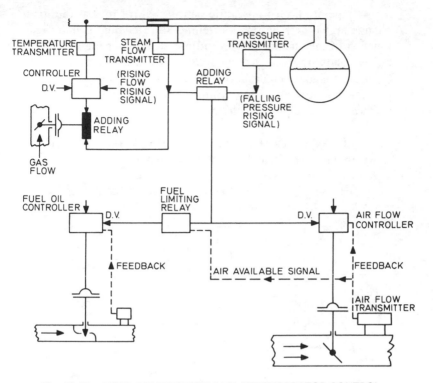

Fig. 13.16 AUTO-COMBUSTION AND ATTEMPERATOR CONTROL SYSTEM

The common signal at the adding relay for pressure and flow adjusts the desired values of both air and oil flow systems. As both these loops are independent sub-groups of a closed nature they will keep changing respective flow rates until the desired value (or set value) is reached. Under normal operation steam flow signals are in advance of steam pressure signals and this anticipatory action is good control practice. The fuel limiting relay in the oil sub-loop (fuel-air ratio control) receives the feedback signal from the air flow rate and compares the required fuel rate with air availability. Fuel rate cannot be increased until an increased air flow signal allows the alteration of the desired value setting at the fuel controller.

Attemperation also has anticipatory control in that the load change signal acts before the temperature change signal. The function of the attemperator is to limit steam temperature at low boiler loads, there is a consistent relation between the two. Command signal choice between fuel pressure, fuel flow, fuel flow/steam flow etc. show significant variations in response characteristics and research still proceeds.

(2) AUXILIARY BOILER AUTOMATIC FUEL AND CONTROL SYSTEM
   Refer to Fig. 13.17 for the **lighting sequence.**

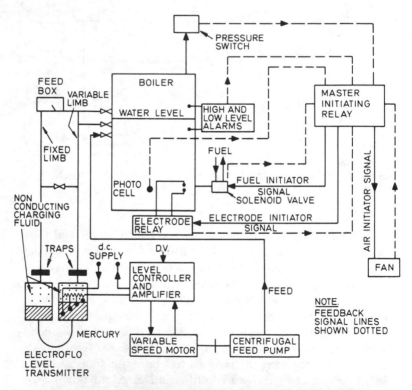

Fig. 13.17 AUXILIARY AUTOMATIC BOILER CONTROL SYSTEM

1. The pressure switch initiates the start of the cycle. The switch is often arranged to cut in at about 1 bar below the working pressure and cut out at about 1/5 bar above the working pressure (this differential is adjustable).

2. The master initiating relay now allows "air-on". The air feedback confirms "air-on" and allows a 30 second time delay to proceed.

3. The master now allows the arc to be struck by the electrode relay. The "arc made" feedback signal allows a 3 second time delay to proceed.

4. The master now allows the fuel initiating signal to proceed. The solenoid valve allows fuel on to the burner. The "fuel on" feedback signal allows a 5 second time delay to proceed (this may be preceded by a fuel heating sequence for boiler oils).

5. The master now examines the photo electric cell. If in order the cycle is complete, if not then fuel is shut off, an alarm bell rings and the cycle is repeated.

Refer to Fig. 13.17 for **emergency devices:**

Obviously failure of any item in the above sequential cycle causes shut down and alarm. In addition the following apply:

a. High or low water levels initiate alarms and allow the master to interrupt and shut down the sequential system.

b. Water level is controlled by an electroflo type of feed regulator and controller. Sequential level resistors are immersed in conducting mercury or non-conducting fluid, so deciding pump speed by variable limb level. The fixed limb level passes over a weir in the feed box.

(3) DRUM LEVEL CONTROLLER (FEED REGULATOR)

Robot feed regulators are proportional controllers (single term) working on a fairly sensitive proportional band. Due to drum contents "swell" and "shrinkage" during manoeuvring load changes, the action is temporarily in the wrong direction. This wrong action is very severe due to the narrow bandwidth and hand operation of feed checks was often necessary. Proportional action is made less sensitive, this reduces the severity of the short term wrong way action but introduces offset. Offset is got rid of by the addition of integral action, *i.e.* the control is two term for the single element action.

A two element action is obtained by incorporating a steam flow measurement to reduce severe feed flow variations when manoeuv-

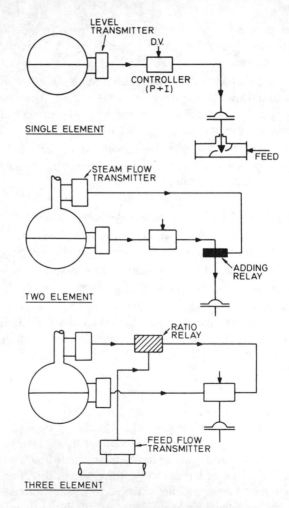

Fig. 13.18 FEED REGULATORS

ring. This signal would give an anticipatory action, which is usually desirable in all control systems. The level signal would act as a trimmer and has a wide proportional band so as not to affect the system during swell and shrinkage.

Three element control gives the highest value of performance. Feed flow is compared to steam flow for the correct 1:1 ratio. If the ratio is incorrect then an out of balance signal is given to the controller. Drum level again acts as a trimming device on a wide proportional band with integral action. Single, two and three element actions are illustrated in Fig. 13.18.

(4) VISCOSITY CONTROL

Refer to Fig. 13.19:

A continuous sample of oil is passed across a capillary tube. The measurement of viscosity has been considered previously (Chapters 5 and 12). Flow is laminar in the tube so that viscosity is directly proportional to pressure drop. Pressure difference is sensed by dp cell transmitter and the signal passed to a controller and recorder. The controller if supplied by air can transmit a direct power signal

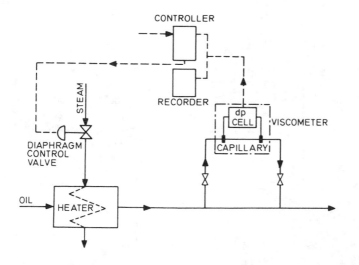

Fig. 13.19 VISCOSITY CONTROL

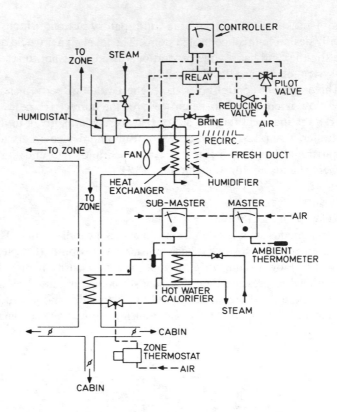

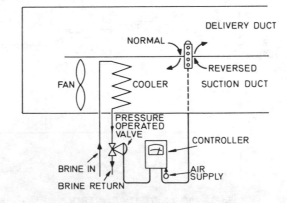

Fig. 13.20 REFRIGERATION CONTROL

to operate a diaphragm control valve. This valve controls steam input to an oil fuel heater. P control is generally adequate, rate and/or reset are easily added. The sensor has been described previously (Fig. 5.3) and so has the controller (Fig. 12.12).

(5) REFRIGERATION CONTROL

Refrigeration and air conditioning utilise some considerable degree of control. Two examples are now given (Fig. 13.20):

Air Conditioning (upper sketch)

Air (fresh and recirculated) is control sprayed with steam to fix humidity. Air is heated (steam grid) or cooled (brine grid) with steam (or brine) quantity controlled by temperature. The air is now passed to the various zones where sub-units adjust the air temperature to the thermostat setting of the zone. This is achieved by controlling the steam supply to a calorifier.

Refrigeration Chamber (lower sketch)

The brine quantity, for adjusting air temperature, is controlled irrespective of fan direction (suction or delivery) with controller bulb in bypass pocket sensing air delivery temperature to chamber.

(6) ALTERNATOR CONTROL

Consider the sketch of Fig. 13.21 and regard this as a main system (A-J) and sub system (K-M).

Main system (load sharing)

One alternator is shown (A) of say a four set installation. Current, voltage, power factor sensors give power computation at relay (B) which is amplified (C). Total electrical load is computed at relay (E), fractioned off to enter relay (D). The two signals at D are compared for load sharing, error signal triggers (F) and thyristor switches (G) for increase/decrease speed signal to governor controller (H) with feedback loop to alternator input power. Total computed power is fed to relay (I) which functions to start another alternator at say 75% maximum rated load of those alternators in operation. The signal is also fed to stop relay (J) which is arranged to shut down an alternator when computed load is say 60%, 40% and 20% of maximum rated load for the four alternator unit.

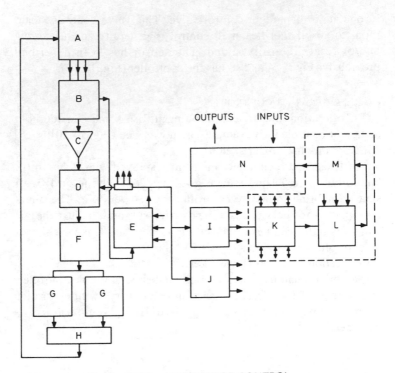

Fig. 13.21   ALTERNATOR CONTROL

Sub-system (alternator start)

This particular sub-system is shown dotted on Fig. 13.21. Obviously a start signal for another required alternator will need a pre-start routine relay (K). This would initiate air, lubrication, cooling, input power, etc., and start the alternator up to running speed. Relay (L) will arrange synchronisation (voltage, frequency, phase) and initiate a signal to close circuit breaker (M).

The main control switchboard (N) would be arranged to handle all monitored inputs from individual alternators. Standard protection from engine faults would be provided. Normal electrical protection is required, for example, reverse power trips, overload alarm (105%) and trip, preferential tripping, etc. These could be regarded as module systems within a particular sub-system. Obviously a shut down arrangement sub-system, similar to *K-M*, is required in which individual alternator off loading takes place and circuit breaker opening at 10% full load occurs.

(7) BUTTERWORTH HEATING CONTROL

On Fig. 13.22, P_1 and P_3 are pressure sensors, P_2 pressure controller; T_1 temperature sensor and T_2 temperature controller; L_1 level sensor and L_2 level controller; R_1 and R_2 relays; A_1 and A_2 actuators. Signals of pressure and temperature enter R_1 and the lower value is passed on to R_2. Control is inherently on temperature with pressure over-ride. Signals from R_1 and the water-pressure sensor enter R_2 and the lower value is passed on to A_1. This provides protection against water supply failure. Conden-

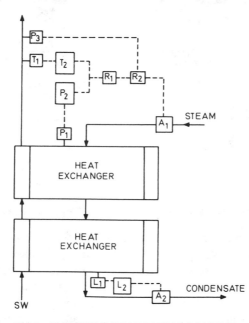

Fig. 13.22 BUTTERWORTH HEATING CONTROL

sate level is controlled as shown. The system can be pneumatic, electronic or a combination of the two. The control problem is essentially the difficulty of handling large quantities of water, with high velocities, maintaining close temperature control and controlling one variable by means of another variable. The system shown has utilised cascade control principles, effectively temperature master reset to pressure slave, the latter being a pressure control system.

(8) OILY-WATER SEPARATOR INTERFACE LEVEL CONTROL

The level sensor has been described in Chapter 3. The control system of Fig. 13.23 utilises two probes with the lower probe (shown) giving a balanced electrical bridge in water and the upper probe (not shown) giving balance in oil or air.

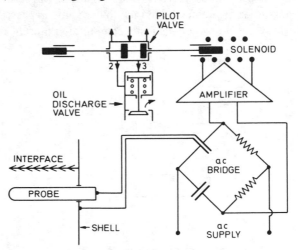

Fig. 13.23   OILY-WATER SEPARATOR INTERFACE LEVEL CONTROL

With the pump started and supplying water to the separator to rise to the lower probe level, the bridge is balanced and the solenoid de-energised. When water rises to the upper probe its bridge is unbalanced and the output signal is amplified which energises a "left-hand" solenoid (not shown) which moves the pilot valve to the left. This allows clean water pressure to pass (from 1) to close the oil discharge valve (through 2). Shell pressure rises and a spring loaded water discharge valve is opened. As oil build up occurs the oil-water interface moves down de-energises the left hand solenoid and then energises the "right hand" solenoid, the pilot valve moves right (as shown) and water pressure opens the oil discharge valve (through 3) and the water discharge valve closes. Each probe and valve has a signal indicator lamp and an alarm bell operates when the lower probe bridge is unbalanced. A third probe at a low level can be arranged to cut out the pump if oil falls to that point.

(9) CONTROLLABLE PITCH PROPELLER

Use of these propellers has increased with the greater use of unidirectional gas turbine and multi-diesel drives and bridge control. Engine room (or bridge) signal is fed to a torque-speed selector which fixes engine speed and propeller pitch - feedbacks apply from each. Consider Fig. 13.24.

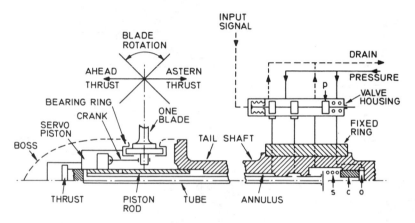

Fig. 13.24 CONTROLLABLE PITCH PROPELLER

The input fluid signal acts on the diaphragm in the valve housing and directs pressure oil via one piston valve through the tube to one side (left) of the servo piston or via the other piston valve outside the tube (in the annulus) to the other side (right) of the servo piston. Movement of the servo piston, through a crank pin ring and sliding blocks rotates blades and varies pitch.

The feedback restoring signal, to restore piston valves to the neutral position at correct pitch position, is dependent on spring(s) force (*i.e.* servo piston position) which acts to vary the orifice (o) by control piston (c) so fixing feedback pressure loading on the pilot valve (p) in the valve housing. The central part of the tailshaft includes a shaft coupling and pitch lock device (not shown).

(10) AUTO STEERING (BLOCK DIAGRAM)

The ship steering gear (on auto pilot) utilises classic control principles best illustrated by a block diagram.

Fig. 13.25 shows a block diagram for auto steering. The controller will be three term with adjustment for beam sea (or wind) and dead band operation to reduce response to small random signals. Both rudder and ship are acted upon by external forces.

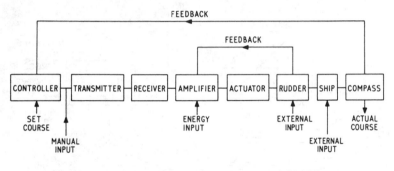

Fig. 13.25 AUTO STEERING (BLOCK DIAGRAM)

(11) 'HEAVY' FUEL OIL SEPARATION

Consider the circuit as sketched (Fig. 13.26). The self cleaning purifier has no gravity disc; sludge and water collect on the bowl periphery. When the water build up leads to some discharging with the clean oil it is sensed (capacitance interface detection) at the transducer which signals the microprocessor. If the time since the last sludge discharge exceeds a pre-set value the sludge and water is discharged from the bowl, otherwise water only is discharged through the water valve. Essentially as a separator but works also as a purifier to discharge water via a paring disc when the water valve is opened.

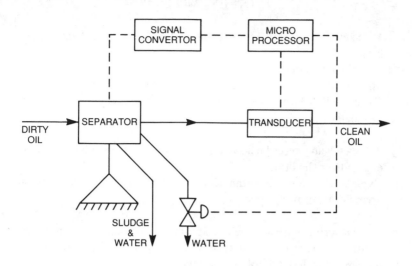

Fig. 13.26 'HEAVY' FUEL OIL SEPARATION

TEST EXAMPLES 13

1.   Sketch and describe a fully automatic, oil fired, packaged steam boiler.
Explain how it operates.
State what attention is needed to ensure safe operation.
Give two advantages possessed over conventional boiler installations.

2.   Sketch and describe a system of control for manoeuvring a main engine from the bridge.
Explain how control is transferred to the engine or control room upon failure of bridge control.

3.   Describe with a block diagram, an automatic combustion control system, with particular reference to the methods of measuring each of the following data:

  (a) pressure,
  (b) level,
  (c) flow,
  (d) temperature.

4.   Define the terms:

  (a) cascade control,
  (b) split level control.

Discuss the application of such principles within a description, utilising sketches, of an automatically controlled lubricating oil system and an automatically controlled cooling water system of the type used in auxiliary diesel driven generators.

CHAPTER 14

KINETIC CONTROL SYSTEMS

KINETIC CONTROL SYSTEM

A control system, the purpose of which is to control the displacement, or the velocity, or the acceleration, or any higher time derivative of the position of the controlled device.

(It should be noted that forces and torques are involved in the above definition).

SERVO-MECHANISM

An automatic monitored kinetic control system which includes a power amplifier in the main forward path. (Includes continuous, discontinuous, on-off, multi-step, etc. actions).

POSITION SYSTEMS

The control of position (displacement) in a system (linear or angular).

POSITION CONTROL SERVO-MECHANISM (dc)

Consider the electrical example of Fig. 14.1 in which the output servo-motor shaft is required to follow the input shaft. As long as there is a difference of angular position between these shafts, measured by toroidal potentiometers, a difference of potential will cause current to flow in the required direction through the amplifier and servo-motor shunt field. The armature of the servo-motor carries a constant voltage (stability with large ballast resistor R) and hence a torque and motor shaft rotation occurs as soon as the field is excited.

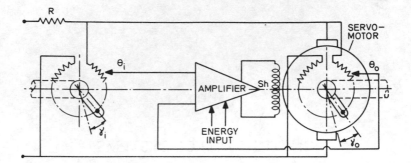

Fig. 14.1 POSITION CONTROL SERVO-MECHANISM (dc)

(A constantly excited motor series field can additionally be connected across the mains if the inertia of the servo-motor is high). As an alternative shunt field current can be arranged to be constant and armature voltage varied.The system as considered however has the true relationship that torque is proportional to actuating signal and is generally independent of rotational speed. The principle is effectively used in practical electrical ship steering gears. (Ward-Leonard system). Reference should also be made to Fig. 14.2 the simple block diagram of the system. Whilst a detailed analysis of the dynamics of such a control system is given later, at this stage it is clear that output torque is proportional to deviation, i.e. proportional action. Field control is used for small powers and armature control for higher powers, the latter utilising a series ballast resistor (dc) or rectifier from ac supply.

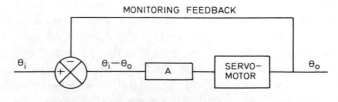

Fig. 14.2 BLOCK DIAGRAM (ONE TERM POSITION)

SYSTEM RESPONSE

The effects of energy transfer including inertia, friction, etc., need to be considered whether mechanical or electrical, etc. Descriptive analysis is often presented electrically for any system due to ease of diagram circuitry and the mechanical, etc., equivalents can then be simply derived.

Consider a *step input* to the servo-mechanism of Fig. 14.1. At the instant of applying input, maximum deviation exists, and the servo motor first accelerates rapidly; when deviation ceases the motor stops. This is an ideal situation because in practice inertia of the motor causes *overshoot*, reverse current and rotation, with oscillation. Such oscillation would continue but for the frictional effects, static and viscous, which damp out oscillation.

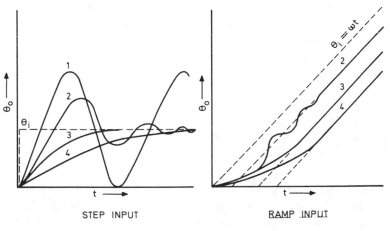

Fig. 14.3 SYSTEM RESPONSE

Referring to Fig. 14.3 for the step input θ_i, curve one represents the undamped oscillation (of natural frequency ω_n) response at output θ_0. Curve two represents light damping (damping factor $k < \omega_n$), curve three critical damping (minimum time without oscillation to equilibrium $k = \omega_n$) and curve four heavy damping, *i.e.* aperiodic ($k > \omega_n$). As a first assumption viscous friction is assumed to account for all frictional effects, with resisting friction force (or torque) proportional to velocity. Damping, to prevent overshoot, limits amplifier gain and speed response.

Also given in Fig. 14.3 is output response θ_0 to *ramp input* θ_i for the three types of damped condition. θ_0 does not equal θ_i in the steady state, *i.e.* error with output lagging input. This is a position lag (offset) due to velocity (rate) which is termed a *velocity misalignment* (friction proportional to velocity). A ramp input of position equals a step input of velocity.

OVERSHOOT

This occurs with proportional control and is due to inertia effects. It can be reduced by any of three methods:

Friction damping.

Friction is non-linear, regarded in two components, static (Coulomb) and viscous, the former giving steady state error and usually neglected in simple analysis. Viscous friction is proportional to velocity and gives damping due to absorption of kinetic energy. Such damping, when utilised, is achieved by increasing load torque as velocity is increased. Damping devices employing viscous friction are not often used because of the following disadvantages: increased response time to achieve steady state, increased losses, increased energy input with a larger velocity misalignment to produce this input. Viscous friction, and static friction, are of course always naturally present to an extent.

Stabilising feedback

Modifying feedforward or feedback is to minimise any tendency to oscillate. In principle the object is to decrease servo-motor output torque as speed increases by tachogenerator feedback, which is preferable to viscous friction damping as the effect is more linear and no extra energy input is required.

Refer to Fig. 14.4 in which the relevant part of the loop is shown dotted, also shown is a gearing (G) and load system (L). Stabilising feedback from the tachogenerator (T_1), proportional to velocity, reduces the voltage input error signal, proportional to deviation, hence the amplifier input is reduced. No input will exist before the shafts are aligned due to the tachogenerator feedback voltage reduction. Inertia will move the output shaft and reverse amplifier output will provide a braking torque to bring the motor to rest. For a ramp function (linear displacement variation, constant velocity) velocity misalignment exists because with the input shaft stopped in position a zero input voltage to the amplifier is required before shaft alignment. This requires an equal and opposite input voltage to the tachometer output voltage available in the main feedback circuit, to balance. Motor and tachometer unit is often called a velodyne.

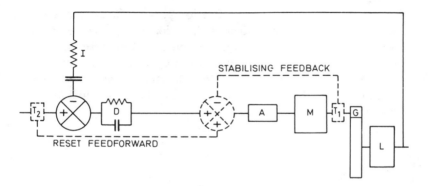

Fig. 14.4   BLOCK DIAGRAM (THREE TERM, POSITION)

Rate network

From the above it is obvious that a voltage proportional to deviation plus a voltage proportional to rate of change of deviation (*i.e.* velocity) is required, which is $P + D$ action. The derivative (rate) circuit has been covered in detail for process control. It is indicated at D and is a phase advance whose voltage requires extra amplifier gain.

For a step input, peak overshoot, settling time and rise time tolerances may be specified as a given percentage of the step change. With a ramp function it is necessary to specify allowable velocity misalignment (function of rate of input change) as a given percentage of maximum input velocity (rate of ramp input

OFFSET

This has been discussed in process control and is a characteristic of proportional control at different loads. Obviously in the position system being discussed it is velocity misalignment. This is a steady state position error due to viscous torque (proportional to velocity) additive to load so that offset must occur. It can be eliminated by one of two methods, the former preferential:

Reset network

A voltage proportional to deviation plus a voltage increasing with time at a rate proportional to deviation is required, *i.e. P + I* action. This has been discussed previously and the reset network (*I*) is shown on Fig. 14.4. The capacitor stores enough delayed charge to feed the amplifier to reduce deviation to zero against the extra loading of frictional effect. It can assist against initial inertia and friction loading.

Reset feedforward

A second identical tachogenerator (T_2) is driven off the input shaft and supplies a feedforward additive voltage proportional to speed. In the steady state for a ramp function the feedforward from T_2 balances the feedback from T_1. Hence no velocity misalignment, offset removed. When the input shaft stops $(T_2$ zero output) the following output shaft rotation provides the usual stabilising feedback rate action from T_1 to reduce overshoot.

POSITION CONTROL SERVO-MECHANISM (ac)

Generally limited to small powers and although cheap have a low start torque and relatively low performance. Essentially two categories of ac system exist, *i.e.* demodulator-modulator and all ac. The principles involved in the former have been discussed in Chapter 7, with application to electronic controllers in Chapters 11 and 12.

Refer now to Fig. 14.5

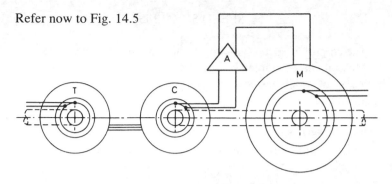

Fig. 14.5 POSITION CONTROL SERVO-MECHANISM (ac)

The input shaft is connected to the single phase ac rotor of the transmission (*T*) and the output shaft to a similar rotor of the controller (*C*) whose excitation is amplified (*A*) for supply to the main drive servo-motor (*M*) to bring the shafts into alignment. Three phase stators of transmitter and controller are directly connected. A null point of relative rotor positions exists, variation from which gives a proportional voltage in one phase or the other. This is *not* a synchro transmission link because the output shaft rotor is *not* mains excited.

The two transducers are three phase induction motors (as synchros-magslips) and usually the main output servo-motor is two phase (one fixed, one control).

Stabilising feedback can be used with an ac tachogenerator. The stator is wound with an input reference field which acts as excitation and an input field wound at right angles. Rotor cutting of reference field induces an emf in the output field proportional to speed and in phase and frequency with input signal.

HYDRAULIC POSITION CONTROL SERVO-MECHANISM

Hydraulic position control has many applications. A typical example would be an electro-hydraulic steering gear. The control function acts to vary plunger travel, often by radially-displaced or "swash plate" piston operated devices, and so give a variable delivery pump. Linear or angular operation to any position control system, particularly hydraulic, is easily arranged.

SPEED SYSTEMS

The control of speed (velocity), linear or angular, in a system.

Most of the principles of speed control are applicable to position control which has been covered so that only a brief analysis is required.

SPEED CONTROL SERVO-MECHANISMS (dc)

The essential equations relating to dc motors are:

$$I \propto \Phi I_a$$

where T is output torque, Φ flux and I_a armature current.

$$N \propto \frac{V - I_a R_a}{\Phi}$$

where N is speed, V applied voltage and R_a armature resistance.

$$P \propto V I_a - I_a^2 R_a$$

where P is output power.

Obviously speed control can be effected by varying armature voltage or field flux, the former being shown on Fig. 14.6.

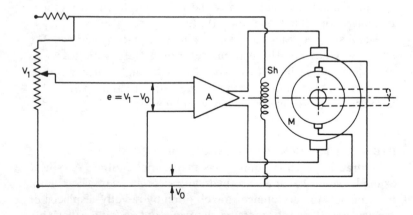

Fig. 14.6 SPEED CONTROL SERVO-MECHANISM (dc)

A speed set voltage from the input potentiometer has an input volts signal V_1. Monitored feedback from the tachogenerator gives a voltage (V_0) proportional to output shaft speed. The error signal e = $V_1 - V_0$ is amplified and fed to main drive motor.

A derivative (rate) and integral (reset) circuit could be added. Rate of velocity change is acceleration. Gyros measure acceleration in many control systems.

SPEED CONTROL (ac)

The actual control unit requires modulation, demodulation and compensation networks, as discussed previously and devices are readily available utilising ac amplifiers. However the servo-motor itself poses real difficulties as the torque-speed characteristic is non-linear. High resistance rotors and thyristor circuitry gives improvements, with higher power induction and synchronous motors, but has disadvantages. In general, except for low power position systems and ac amplifier control units, the all ac system is not greatly used as yet. However a mixed ac-dc arrangement for speed control has many advantages. The power amplification-conversion device utilised to what is essentially a dc drive system from ac supply is an important consideration. Vacuum valves (thermionic) are limited to very small powers. Magnetic amplifiers, with main and field solid state rectifiers, are reliable and can be used directly with three phase for larger powers or as field controllers in a motor generator set. The motor generator set of the Ward-Leonard type is often used for very high powers but is a large unit, subject to time delays in operation.

WARD-LEONARD SPEED CONTROL

Consider the arrangement shown diagrammatically in Fig. 14.7:

Terminology and principle of operation should be clear from the sketch. The input motor (DM) is a constant speed induction motor with standard dc generator (G) and servo-motor (M), the latter with a constant excitation field. Generated supply voltage to the servo-motor depends on error signal between desired and measured speed values, as proportional signal voltages, which is

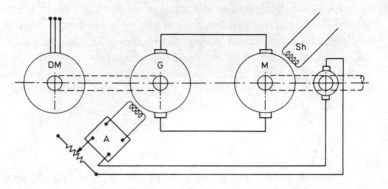

Fig. 14.7 WARD-LEONARD SPEED CONTROL

amplified and fed to the generator control field. For small powers a valve or transistor amplifier could be used with extension to medium powers with a magnetic amplifier. High gain and stability from drift can be obtained utilising such ac amplifiers suitably modulated and demodulated. For high powers it is usually necessary to replace these amplifiers with either a rotating amplifier – auxiliary generator (exciter) or amplidyne (metadyne, cross field dc generator) – or modern thyristor control amplifier.

Controlled rectifier units can also be used. The mercury arc rectifier and ignitron can be used for large powers but are generally limited to specialist applications. Thyratron devices, with transformer coupling to anode and grid, have been successfully used. Thyristors are being increasingly used either directly or as field control devices.

THYRISTOR SPEED CONTROL

Consider the arrangement shown in Fig. 14.8:

The sketch has the familiar layout used previously, field control is by thyristor. The bridge circuit consists of two rectifiers C and D and two controlled rectifiers (thyristors) at A and B. The gate of each thyristor is triggered by pulses from P representative of error speed input signal. Alternating current supply is rectified for output to field by passage through B and C on one cycle and D and

A on the other cycle. Full wave rectification with thyristor trigger control on each cycle. Adequate overload protection is required in the bridge circuit.

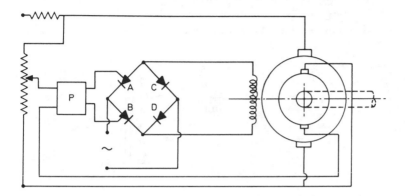

Fig. 14.8   THYRISTOR SPEED CONTROL

GOVERNOR SYSTEMS

Many engines, turbine and reciprocating, are still fitted with trip devices to allow full energy supply under normal conditions. If revolutions rise about 5% above normal the energy supply is cut off until normal conditions are restored - lock out occurs at about 15% excess which can only be unlocked by hand. Aspinall types come into this category.

Smaller engines, such as electric generator drive often use centrifugal governors based on the Watt principle, the Hartnell governor is typical. Control is essentially proportional action with sensed output (rotation speed) controlling energy input and offset (exemplified by no load to full load speed droop) occurs.

Modern engine governors are isochronous devices but reset action is applied to eliminate hunting. One such $P + I$ governor has already been described in Chapter 12.

Two designs will now be considered namely Mechanical-Hydraulic (similar in principle to Fig. 12.2) and Electrical-Hydraulic (utilising principles discussed previously in this Chapter).

MECHANICAL-HYDRAULIC SPEED CONTROL SERVO-MECHANISM

Many engine units employ these servo devices incorporating in-built safety for such as oil or water failure with IC engines. The design now considered is similar in principal to that of Fig. 12.2.

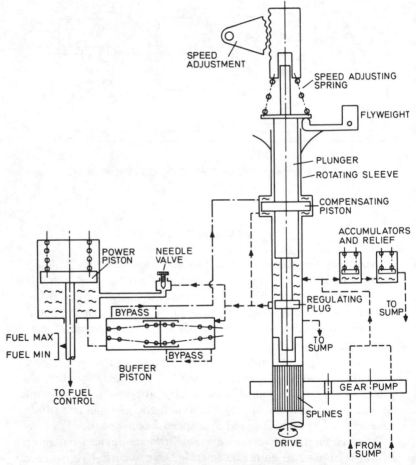

Fig. 14.9 MECHANICAL-HYDRAULIC SPEED CONTROL
SERVO-MECHANISM

Refer to Fig. 14.9:

With the engine running at constant speed under a steady load the up-force due to centrifugal force from the flyweights is balanced by the down-force of the speeder spring. The plunger is central with the regulating plug covering the regulating ports in the sleeve. The plunger moves vertically but does not rotate and the opposite applies to the sleeve. The power piston is stationary and the buffer piston central under these conditions.

Consider a load *increase* on the engine for which condition Fig. 14.9 is applicable. Speed reduces and the plunger moves down with pressure oil flow to the right of the buffer piston, which moves left. The power piston will move up and admit more fuel to the engine. Pressure oils also acts on the compensating piston under side which will exceed the pressure on this piston upper side so that the plunger will be restored up. The power piston will now stop. As engine speed returns to normal oil is leaking through the needle valve, to restore equal pressures on each side of the buffer piston and compensating piston. The buffer piston is returned to mid-position by the springs. This gradually reduces the up-force on the compensating piston but the increasing engine speed is also increasing this up-force due to centrifugal force. The compensating piston is designed to be balanced gradually so that the rate of leakage at the needle valve (unloading) equals the rate of loading due to extra centrifugal force caused by higher engine revolutions. The engine will now run at normal speed but with increased load and higher fuel setting. The needle valve should be screwed in sufficiently to prevent hunting but without making the operation sluggish. Oil pressure is shown dotted with the slightly lower pressure chain dotted.

The bypass arrangement ensures that for a large speed change the power piston only moves as far as the bypass. Pressure oil flows directly to the power cylinder without further increasing the pressure differential on the compensating piston. After sufficient governor movement and speed return to near normal the differential pressure acts as usual.

In the event of a large load decrease the power piston is at fuel minimum and blocks the needle valve connection. This gives a higher speed setting than normal and reduces a tendency to under speed.

ELECTRICAL HYDRAULIC SPEED CONTROL SERVO-MECHANISM

Rotational speed is sensed by tachogenerator, or as shown in Fig. 14.10 by ac alternator, with frequency pulses converted in the rectifier to dc voltage proportional to speed. Set value is applied to the controller and the two input voltages, opposite in polarity, are compared. Error signal (if any) is amplified, converted to hydraulic signal, and operates to alter energy supply through a servo-motor. To reduce hunting, and offset, the controller has reset (integral) action via a feedback loop. The unit is also anticipatory *i.e.* load changes are fed back to the controller to amend energy input *before* speed change occurs – speed control is virtually a fine trimming operation.

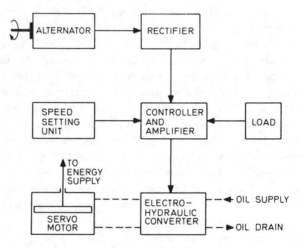

Fig. 14.10   ELECTRICAL-HYDRAULIC SPEED CONTROL
SERVO-MECHANISM

MATHEMATICAL ASPECTS

Fig. 14.11 is introduced now to link illustrative block diagrams (as Fig. 14.2) and the analyses of transfer functions for open and closed control loops (Chapter 15). It is a position control servo-mechanism with unity feedback. Torque proportional to deviation, friction torque proportional to (angular) velocity.

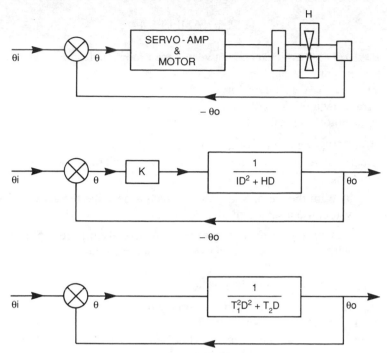

Fig. 14.11 BLOCK DIAGRAM/TRANSFER FUNCTION

Torque applied to load $(I\alpha) = K\theta - H\omega$

$$K\theta = I\frac{d^2\theta_0{}^2}{dt^2} + H\frac{d\theta}{dt}$$

$$= \theta_0 (ID^2 + HD)$$

Open loop transfer function

$$\frac{\theta_0}{\theta} = \frac{K}{ID^2 + HD} = \frac{1}{T_1{}^2D^2 + T_2D}$$

Closed loop transfer function

$$\frac{\theta_0}{\theta_i} = \frac{(T_1{}^2D^2 + T_2D)^{-1}}{\{1 + (T_1{}^2D^2 + T_2D)^{-1}\}} = \frac{1}{(T_1{}^2D^2 + T_2D + 1)}$$

where $T_1 = \sqrt{\dfrac{I}{K}}$ and $T_2 = \dfrac{H}{K}$ are time constants

See now Pages 261 and 262.

TEST EXAMPLES 14

1. Explain the difference between "open loop" and "closed loop" systems of control.
Draw a circuit diagram for a system in which the Ward-Leonard arrangement with feedback control is used to regulate the speed of a dc motor and explain the mode of operation.

2. (a) Sketch a clearly-labelled circuit diagram for a simple electrical remote position control servo-mechanism with zero damping.

(b) With the aid of wave form sketches describe the action of the system when subjected to a step input.

(c) Compare any advantages and disadvantages of the following methods of damping such as a servo-mechanism:
   i. Viscous friction,
   ii. Output velocity feedback.

3. Describe a thyristor control arrangement for the speed control of a large electrical fan. Show how a zener diode can be used to stabilise the voltage supply to load and include the necessary protection to safeguard the diode against overload.

4. (a) Distinguish between the following control system terms:
   i. Error   ii. Offset
   iii. Monitored feedback   iv. Deadband

(b) For a remote position control servo-mechanism:
   i. State the effect of adding integral action,
   ii. Describe velocity feedback damping,
   iii. Sketch waveforms to illustrate response to step input.

CHAPTER 15

CONTROL SYSTEM ANALYSIS

This subject is generally complex and the objective of this chapter is to introduce the basic principles so as to allow an initial appreciation which could be further developed, if required, at a later stage. Consideration is given to the systems approach, the order of linear systems, performance of systems, component interaction and adjustment.

THE SYSTEMS APPROACH

SYSTEM

Capable of many and varied definitions. A general definition could be: a functional assembly with components linked in an organised way and affected by being within, and changed if removed from, the boundary.

May exist as sub-systems within a larger system so that a hierarchy exists, for example the biological cell, within the heart, of an animal, within a social human system, of a universe.

STATE

A system may be discrete, *i.e.* exist in one only clearly defined state at a given time, or may be continuous in change. Can be deterministic in operating to a fixed sequence, or probabilistic with random change or subject to external influence. A closed system always tends to seek equilibrium and ideally has no energy transfer with surroundings outside the boundary whereas an open system tends to approach a steady state of balance with the surrounding environment. The **black box** philosophy is applicable, internal form unknown and sealed, with the only factors of interest being the output and input variables and their relation.

SYSTEMS APPROACH

There are barriers, sometimes artificial, between specialist subjects which are in many cases being eroded by the unifying theme of technology now being adopted through a systems approach to problem analysis and solution. Mathematical models can be used to describe a variety of systems ranging from mechanical control to organisational man management. Analogies between electrical and mechanical systems can easily be illustrated but the principles established are often more widely ranging. The aspect is too broad for detailed consideration at this stage but the technique of moving from the particular (say engineering), by analogue and commonality, to the general system is a desirable aim. It is therefore usual to demonstrate a systems approach to an engineering situation and to use this as a vehicle to suggest that this systems approach applies throughout and a general theory may be attainable.

Essentially the systems approach can be detailed as:

1. Specify aims and objectives of the problem or analysis.

2. Establish system and sub-system boundaries.

3. Devise functional conceptual models of the problem leading to block diagrams with attendant mathematical models (equations) allowing for interaction between component units, feedback analyses, etc.

4. Scale system variables and construct analogue and circuit diagrams.

5. After evaluation and iteration the synthesis can be established and tested for final appraisal.

(The reader is strongly advised at this stage to consider Fig. 16.1A of the next chapter for a systems approach applied to a very basic engineering mechanism, i.e. the simple pendulum and its analogue).

Development to transport, banking, education, manufacturing, community systems, etc., is an essential part of general systems theory and philosophy based on the techniques outlined in principle above.

SYSTEM ORDER

Systems considered in this context are linear, i.e. equations with constant coefficients. In practice non-linearities exist but unless complicated theory is utilised are difficult to analyse. Many cases exist where the effect of non-linearity can be negligible by correct design so that linear theory can be applied.

A systems approach requires a unification between similar quantities and should result in a generalised mathematical model whose equations are applicable for simulation and evaluation.

ANALOGUES

It should be remembered that rate of change of a variable with respect to time can be written d/dt in calculus notation. Thus, for example, translational velocity (rate of change of translational displacement with respect to time) can be written dx/dt; an alternative is $\dot{x}$. Similarly, for example, rotational (angular) acceleration is dw/dt, or $\dot{\omega}$ and as this is the second rate derivative of rotational displacement it also equals $d^2\theta/dt^2$; or $\ddot{\theta}$. Numerous other examples can be quoted.

Variables in common use include force, torque, voltage (drop), pressure (drop), displacement, velocity, acceleration, current, flow rate, etc. *Parameters* include stiffness, damping coefficient, mass, inertia, resistance, capacitance, inductance, etc (see Table 15.1). Consider the following:

$$\textbf{Generalised Impedance } (Z) = \frac{\textbf{Across Variable } (X)}{\textbf{Through Variable } (Y)}$$

Through variables are such as velocity, current and flow rate; across variables are such as force, voltage and pressure; impedance parameters are such as inertia, resistance and capacity. Elements in a system can be classified as dissipative, where $X \propto Y$ (such as resistors); as *storage*, where $\dot{X} \propto Y$ (such as capacity); or as *storage*, where $X \propto \dot{Y}$ (such as inductance). Table 15.1

compares the translational mechanical, electrical and fluid systems. Extension to pneumatic, thermal, and rotational mechanical systems gives a similar result in every case. B is damping coefficient, S' spring stiffness, the dot above variables indicating rate of change, *e.g.* $\dot{v}$ will be acceleration.

Gen.	X	Y	$Z = \dfrac{X}{Y}$	$Z = \dfrac{\dot{X}}{Y}$	$Z = \dfrac{X}{\dot{Y}}$
Mech.	Force	Velocity	Damper $B = \dfrac{F}{v}$	Spring $S' = \dfrac{\dot{F}}{v}$	Mass $m = \dfrac{F}{\dot{v}}$
Elec.	Voltage	Current	Resistor $R = \dfrac{V}{I}$	Capacitor $\dfrac{1}{C} = \dfrac{\dot{V}}{I}$	Inductor $L = \dfrac{V}{\dot{I}}$
Fluid	Pressure	Flow	Resistive $R' = \dfrac{p}{f}$	Capacity $\dfrac{1}{C'} = \dfrac{\dot{p}}{f}$	Inertive $I' = \dfrac{p}{\dot{f}}$

Table 15.1

SYSTEM ORDER

The response of any system, or a component within the system, can be described by a mathematical equation. The order of the equation, and hence the system or component, is fixed by the highest derivative "power". In a mechanical system, for example, velocity is the first derivative of displacement, *i.e* $v = dx/dt$ so that a system with such a term as the highest derivative is classified as a *first order* equation. All such first order systems are defined by having *one* form of energy storage component. Similarly acceleration as the second derivative of displacement, *i.e.* $a = d^2x/dt^2$ is a *second order*. All such second order systems are classified by having *two* forms of energy storage components and *one* form of dissipative energy component. It is often convenient to write D for d/dt in calculus notation (note D^{-1} is a first *integration*).

FIRST ORDER SYSTEMS

Consider the mechanical translation system of Fig. 15.1:

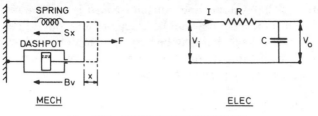

Fig. 15.1 FIRST ORDER SYSTEMS

The constant applied force (F) is resisted by the spring force $(S'x)$ and the dashpot damping force proportional to velocity (Bv). Now $v = dx/dt = Dx$ so that the equation is written:

$$F = S'x + BDx$$

and after re-arrangement this becomes

$$\frac{B}{S'}Dx + x = \frac{F}{S'}$$

which is a typical first order equation, time constant $\tau = \dfrac{B}{S'}$

For the electrical resistance-capacity system shown:
$$\tau DV_0 + V_0 = V_i$$

(in charge terms, $\tau DQ + Q = V_iC$ when DQ is current)

where $\tau = RC$ is a time constant.

For a heat resistance-capacity system:
$$\tau D\theta_E + \theta_E = \theta_F$$

where $\tau = RC$, θ_E and θ_F element and fluid temperatures.

For a fluid restrictor-capacity tank system:
$$\tau Dp_0 + p_0 = p_i$$

(in quantity terms, $\tau Dq + q = p_iC'$ where Dq is flow rate)

("   head   ", $\tau Dh + h = \dot{q}_iR$ " $\tau = $ Area x R)

Such equations can be extended to economic, management, etc. systems and a general equation arising from the above analogous cases may be written down as follows (y any variable):

$$\tau DY + y = bf(t)$$

where τ is the system time constant and b is a constant. It is possible that the input may not be a step or ramp form of constant, such as a dc voltage, but may be varying with time, such as an ac voltage. The $f(t)$ forcing function term is a general way of writing an input function dependent on time, for example sinusoidally, to allow for such variations. If the input is not so varying the right hand side of the equation is a constant, as covered in the analogous cases given above, when y is made up of a constant and a variable, with a system time constant in the solution *i.e.* $y = b(1 - e^{-t/\tau})$. b is usually the gain constant.

TRANSFER FUNCTION

The transfer function of an element is the ratio of its output signal to its input signal.

$$\text{Transfer Function} = \frac{\theta_0}{\theta_i}$$

Its use, together with block diagrams, simplifies analysis using s-plane in place of differential equations. In some cases output is merely amplified or attenuated input, *e.g.* gearbox, whilst in other cases the signals may be in different physical form with different amplitude and phase.

Consider the *RC* network of Fig. 15.1:

$$V_i = IR + V_0$$

$$I = C\frac{dV_0}{dt} = CDV_0$$

$$V_i = RCDV_0 + V_0$$

(this is the first order equation given previously)

$$V_i = V_0(RCD + 1)$$

$$\text{Transfer Function} = \frac{V_0}{V_i} = \frac{1}{1 + \tau D}$$

This result will be characteristic of all such first order equations.

CLOSED LOOP TRANSFER FUNCTION

Figure 15.2 illustrates the usual arrangement. The transfer function of forward elements is G and of feedback element F. Also see Page 108.

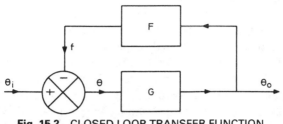

Fig. 15.2 CLOSED LOOP TRANSFER FUNCTION

1. Consider the loop of feedback to be opened and let $F = 1$ (direct feedback):

$$\text{Open Loop Transfer Function } (G) = \frac{\theta_0}{\theta}$$

for forward path elements.

2. Consider closure of the feedback loop and $F = 1$:

$$\text{Closed Loop Transfer Function} = \frac{\theta_0}{\theta_i} = \frac{\theta_0}{\theta + \theta_0}$$

$$= \frac{\theta_0/\theta}{1 + \theta_0/\theta} = \frac{G}{1 + G}$$

3. Consider the system as sketched:

$$f = F\theta_0; \quad \theta = \theta_i - f; \quad \theta_0 = G\theta_0$$

combining these equations gives:

$$\theta_0/G = \theta_i - F\theta_0$$

$$\theta_i = \theta_0 \left(\frac{1 + FG}{G} \right)$$

$$\text{Closed Loop Transfer Function} = \frac{\theta_0}{\theta_i} = \frac{G}{1 + FG}$$

$$\text{Open} \quad " \quad " \quad " \quad = \frac{f}{\theta} = FG$$

1. The open loop transfer function is very useful in stability testing (see later).

2. Note that for direct feedback the closed loop transfer function is (open loop transfer function) divided by (one plus open loop transfer function). Increasing gain G reduces offset.

3. This is a simple example. Practical cases with more involved transfer functions are more complicated and difficult to solve.

The response output θ_0 will depend on the form of the input signal. For elements in series the overall transfer function is the product of the individual transfer functions , assuming there is no interaction between the components *i.e.* $G = G_1 G_2$. There will be an overall gain (or attenuation) and if input is a variable with time there will be a need to algebraically add phase shifts. With elements in parallel $G = G_1 + G_2$. For a given open loop transfer function there is only one closed loop transfer function, hence open loop analysis gives closed loop analysis automatically. This fact is very important as an unstable closed system cannot be measured but by opening the system an analysis is possible allowing stability compensation to be made. The open loop transfer function, independent of where the loop is opened, is FG.

If the system under consideration was itself a component of a control system it could be represented as one block diagram enclosing its transfer function. Such a second order system is shown in Fig. 15.3.

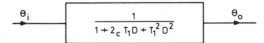

$$\theta_i \rightarrow \boxed{\dfrac{1}{1 + 2_c T_1 D + T_1^2 D^2}} \rightarrow \theta_o$$

Fig. 15.3   TRANSFER FUNCTION BLOCK DIAGRAM

SECOND ORDER SYSTEMS

Such systems have analogues as detailed previously. A review in depth is not required but three examples are given:

1. RCL series electrical network.

Applied volts = Resistor volts + Capacitor volts
+ Inductor volts

$$V = IR + \frac{1}{C} \int I \, dt + L \frac{dI}{dt}$$

$$\frac{dV}{dt} = R \frac{dI}{dt} + \frac{I}{C} + L \frac{d^2I}{dt^2} \quad \text{(after differentiation)}$$

2. Mechanical translational damper, spring, mass assembly.
(See Fig. 15.10).
Applied force = Damping force + Spring force + Inertia force.

$$F = Bv + S'x + ma \quad (B \text{ damping coefficient})$$

$$F = B \frac{dx}{dt} + S'x + m \frac{d^2x}{dt^2} \quad (S' \text{ spring stiffness})$$

$$\frac{d^2x}{dt^2} + 2k \frac{dx}{dt} + \omega_n^2 x = \frac{F}{m}$$

$2k = \dfrac{B}{m}$ where k is the damping factor

$\omega_n^2 = \dfrac{S'}{m}$ where ω_n is the natural (undamped) frequency

$$D^2x + 2c\omega_n \, Dx + \omega_n^2 x = \frac{F}{m}$$

$c = \dfrac{k}{\omega_n}$ where c is a damping ratio

(constant changing is purely for mathematical convenience, m is
the effective mass of the spring and its load).
3. Position control servo-mechanism.
Reference should be made to Fig. 14.1, Chapter 14, showing
this unit. Output drive torque is dependent on inertia, viscous and
stiffness (static and load error) torques.

$$K_1 K_2 K_3(\theta_i - \theta_0) = Ia + H\omega$$

(I is the moment of inertia, H is damping coefficient)

$$\frac{I}{K_1 K_2 K_3} D^2\theta_0 + \frac{H}{K_1 K_2 K_3} D\theta_0 + \theta_0 = \theta_i$$

(Inertia torque is for motor and load, direct drive assumed, viscous torque proportional to angular velocity: K_1, K_2, K_3 are respectively potentiometer bridge, motor torque-current and amplifier *scaling constants*.) Simplifying constants gives:

$$(D^2 + 2c\omega_n D + \omega_n^2)\theta_0 = \omega_n^2\theta_i$$

$$\text{Transfer function} = \frac{1}{T_1^2 D^2 + 2cT_1 D + 1}$$

where $T_1 = 1/\omega_n$ and is periodic time of undamped natural oscillation divided by 2π. For sinusoidal input the transfer function is as above but with $i\omega$ replacing D, *i.e.* if $\theta = Ae^{i\omega t}$ then $D\theta = i\omega\theta$; both D and i are operators.

Note:

A general second order equation, allowing for input variation with time is:

$$(D^2 + 2c\omega_n D + \omega_n^2)y = a\omega_n^2 f(t)$$

where a is a constant, y the variable and $f(t)$ the forcing function.

HIGHER ORDER SYSTEMS

Commonly arise but are not considered here except to note three term control action ($P + I + D$), action factors K_1, K_2, K_3.

$$V = -K_1\left(\theta + \frac{K_2}{K_1}\int\theta dt + \frac{K_3}{K_1}\frac{d\theta}{dt}\right)$$

$$= -K_1\left(1 + \frac{1}{SD} + TD\right)\theta$$

S integral action time, T derivative action time, V controller output. Application to say a position control servo-mechanism gives a transfer function of a third order character obtained by equating the system second order equation in terms of θ_0 to the above equation in which $\theta = \theta_i - \theta_0$.

SYSTEM PERFORMANCE

Ideal response is where output is identical to input command. This cannot be obtained because viscous friction and measure delays result in output lag. Effects such as inertia produce oscillation about the steady state. The mathematical approach gives an estimate of likely performance which can be improved by experimentation rigs, usually electrical analogues. The objective is attainment of desired value fairly quickly and accurately, with stability. Response to simple input signals has already been covered in the text but a quick resumé can now be presented.

Performance results are in two distinct parts:

1. Transient response when the system is responding and may hunt.

2. Steady state response when the transient has died away.

It is necessary to introduce mathematical solutions to equations, especially second order, in the following sections. Such solutions can be readily verified if required but the object in presenting the work is purely to introduce techniques and a general appreciation is all that is required.

STEP INPUT RESPONSE

For a first order system

$$\tau Dy + y = bf(t) \quad \text{is the general equation}$$

$$\tau D\theta_0 + \theta_0 = \theta_i \quad \text{in control terminology}$$

The steady state is a constant step input θ_i. The transient solution is $\theta_0 = -\theta_i e^{-t/\tau}$. The complete solution is $\theta_0 = \theta_i(1 - e^{-t/\tau})$. To "slow down" such a system increase resistance, damping or capacity but decrease stiffness.

Reference to Fig. 15.4 shows an exponential curve. The time constant τ is the time to reach steady state if the initial slope was maintained. Output actually only reaches 63.2% of this value. There is no overshoot. The curve is characteristic of the *LR* electrical "growth" response. The *CR* electrical circuit gives a decay characteristic.

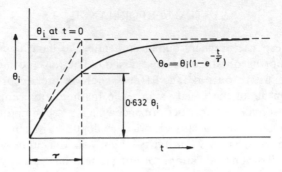

Fig. 15.4 STEP RESPONSE (FIRST ORDER)

For a second order system:

$$(D^2 + 2c\omega_n D + \omega_n^2)y = a\omega_n^2 f(t) \quad \text{general}$$

$$(D^2 + 2c\omega_n D + \omega_n^2)\theta_0 = \omega_n^2\theta_i \quad \text{control}$$

Again the steady state is a constant step input θ_i (after dividing through by ω_n^2). The transient solution is obtained by setting the left hand side to zero and three resulting equations are possible depending on roots obtained. Reference should be made to Fig. 14.3 of Chapter 14 where curves representing the three solution equations are shown. Curve 2 ($c < 1$) is one solution for underdamped and is oscillatory. Curve 3 ($c = 1$) is another solution for critical damping. Curve 4 ($c > 1$) is the third possible solution for overdamped (aperiodic). The oscillatory case illustrates overshoot, a settling time, etc. For the electrical (series) system critical damping occurs when R = $2\sqrt{L/C}$, for mechanical (translational) when B = $2\sqrt{S'm}$.

RAMP INPUT RESPONSE

For a first order system:

The complete solution is $\theta_0 = \omega[t - \tau(1 - e^{-t/\tau})]$. Reference to Fig. 14.3 of Chapter 14 would indicate exponential approach to an inclined line representing steady state θ_0, similar to curve 3 (or 4), starting at τ on the horizontal axis (see Fig. 9.2, Chapter 9).

For a second order system:

Reference to Fig. 14.3 of Chapter 14 illustrates response for underdamped ($c < 1$ curve 2), critical damping ($c = 1$ curve 3) and overdamped ($c > 1$ curve 4). The inclined line of steady state θ_0, as appropriate to the degree of damping, will start at $2cT_1$ on the horizontal axis (see also Fig. 9.2 of Chapter 9).

SINUSOIDAL INPUT RESPONSE

This is the case with the forcing function related to time, i.e. $f(t)$ such as $\theta_i \sin \omega t$ or $\theta_i \cos \omega t$, for say an alternating voltage input; first or second order equation. The system is subjected to varied frequency sine wave inputs and response output is noted in magnitude and phase. Such analysis is also useful for evaluation of higher order systems.

The first order solution to $TD\theta_0 + \theta_0 = \theta_i \cos \omega t$ results in one equation (magnitude) and another equation (phase).

$$\frac{\theta_0}{\theta_i} = \frac{1}{\sqrt{1 + \omega^2 T_1^2}}$$

$$\tan \phi = \omega T_1$$

Consider as example the *RC* network covered previously (Fig. 15.1) with the object of obtaining the steady state sinusoidal response. The method, with sinusoidal type inputs, is to replace operator D by $i\omega$ using the *complex number* notation which is very useful in ac networks or *polar* control plots. Symbol i (sometimes j) is an *operator* which rotates a vector by 90° in an anticlockwise direction, without altering its length, its numerical value is $\sqrt{-1}$; i^2 is 180°, *numerically* -1; i^3 is 270°, numerically $-\sqrt{-1}$, i.e. $-i$; i^4 is 360°, or 0°, i.e. numerically 1. This is illustrated by an **Argand Diagram** of Fig. 15.5 which shows a vector A whose *modulus*, or *amplitude* (length), is 5, i.e. $\sqrt{3^2 + 4^2}$ and *arguement* (phase angle) is tan $^{-1}$ 4/3. The vertical (*Y*) axis is referred to as *imaginary* and the horizontal (*X*) axis as *real* and the number composed of real and imaginary parts is called a complex number. The diagram is often called the "s-plane".

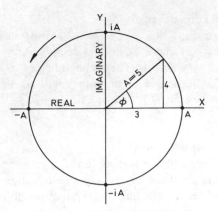

Fig. 15.5 ARGAND DIAGRAM

From the circuit of Fig. 15.1:

$$I = \frac{V_i}{Z} = \frac{V_i}{R - i/\omega C}$$

(Capacitive reactance (Xc) is $i/\omega C$ where $\omega = 2\pi f$, impedance Z.)
($-i$ indicates 270°, *i.e.* a current leading voltage, capacitive effect.)

$$V_0 = \frac{(-i/\omega C)V_i}{R-(i/\omega C)}$$

The next step is to evaluate the transfer function:

$$\frac{V_0}{V_i} = \frac{-i/\omega C}{R - i/\omega C} = \frac{1/i\omega C}{R + 1/i\omega C}$$

$$= \frac{1}{1 + i\omega CR} = \frac{1}{1 + i\omega\tau}$$

This is the typical first order transfer function with D replaced by $i\omega$. This expression gives the modulus, which is termed magnitude ratio (M) in control terms, and argument (phase shift).

$$M = \frac{1}{\sqrt{1^2 + \omega^2\tau^2}}$$

$\phi = \tan^{-1}\omega\tau$ phase of V_0 relative to V_i.

Transfer function second order is $1/(1 + i\omega\tau_1)(1 + i\omega\tau_2)$.

The second order solution is more complicated but shows the same characteristic result of a response of different magnitude and phase to input. Solutions are best illustrated graphically, Fig. 15.6 (also Fig. 9.3, 15.8). Measurement is by a cathode ray oscilloscope or transfer function analyser. Resonant frequency is $1/2\pi\sqrt{LC}$ electrical (series) and $1/2\pi\sqrt{S'/m}$ mechanical (translational).

FREQUENCY RESPONSE ANALYSIS

Referring to Fig. 15.6; the dotted line shows first order response and the full lines are second order response curves for various

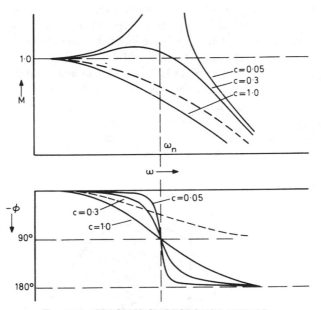

Fig. 15.6   FREQUENCY RESPONSE CURVES

degrees of damping. In the former case magnitude ratio (M the ratio between output and input amplitude) decreases steadily with input frequency and phase lag (ϕ) increases, due to the output being unable to follow input. In the latter case results are similar with a high resonance peak at ω_n for low damping which reduces with increased damping, as does ω_n^2. Lag is inherent in control systems. A frequency response diagram of $M \sim \phi$ with M a logarithmic axis or decibels (dB $= 20 \log_{10} \theta_0/\theta_i$), both effectively the same, is used in open and closed loop analysis. Response diagrams as in Fig. 15.6, but with a logarithmic ω (or ωT_1) axis and a logarithmic M axis (or dB), are similarly used. Such diagrams if used in open loop frequency response analysis are called **Nichols** and **Bode** diagrams respectively. On a Bode diagram, for stability, then i. when dB are zero ϕ should be under $-180°$ (*e.g.* $-135°$) for a positive phase margin and ii. when ϕ is $-180°$ dB should be negative for a positive gain margin.

STABILITY RESPONSE

The main aim of a frequency response test is to assess stability. One common method is to open the feedback loop and inject a small sinusoidal constant magnitude input signal (θ) to the forward path elements only and obtain a polar plot of this open loop frequency response. The input is usually made unity and the polar plot is a **Nyquist** diagram obtained by measuring magnitude ratio and phase angle of output for increasing values of frequency from zero to infinity. M and ϕ could of course be calculated but this is obviously pointless, at least at this stage. A typical Nyquist diagram (open loop polar plot) is shown in Fig. 15.7.

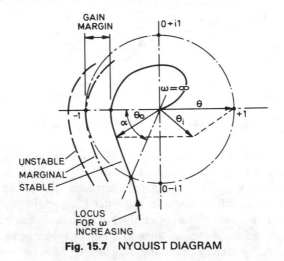

Fig. 15.7 NYQUIST DIAGRAM

The *Nyquist stability criteria* is that the *closed loop system* is stable if the *open loop* frequency response locus traced from w = 0 to $\omega = \infty$ does not enclose (pass to the left of) the point (−1, 0). The locus of θ_0 for ω increasing can be seen to be stable from the diagram. The marginal state through (−1, 0) is shown, with instability beyond that. The circle is for unity gain ($M = 1$). Locus curves starting at various gain values on an extended axis to the right (where $\omega = o$), can be plotted for fixed gain with increasing values of ω to determine correct gain for complete stability.

(Also shown is a vector sum $\theta_i = \theta_0 + \theta$ being the input signal to give $\theta = 1$; this gives indication of *closed loop* response but is not part of the normal Nyquist criteria).

If $\theta_0 = 1$ and $\phi = 180°$ for a particular frequency ω, *i.e.* tip of θ_0 locus vector is point (−1, 0) then θ_i to a closed loop system is zero. This is unity feedback (positive) and slight change of θ_i will cause oscillation which would grow with increased amplitude oscillations for open loop gain over unity. This is the basis of the Nyquist criteria.

Phase margin (α), as shown, should exceed 30° (usual range 30°–60°) and *gain margin* (sometimes expressed in dB) should exceed 0·3 (usual range 0·3 to 0·6) to ensure stability. The curve of Fig. 15.7 is a typical fourth order.

Illustrative response curves are shown in Fig. 15.8 ($M = 1$); a is typical of a first order system, b second order, c is combination of a and b (third order effectively). Curve a is typical of a passive *LR* circuit response and is a 4th quadrant semicircle with output lagging input and amplitude decreasing as ω increases. (A passive *CR* circuit is plotted in the 1st quadrant with output leading input and amplitude increasing as ω increases.) The first and second order systems sketched cannot be made unstable with an increase in gain factor but the third order system can (as can the fourth order of Fig. 15.7). In such a case feedback from a tachogenerator would induce oscillation and increasing amplitude, i.e. instability. This can be prevented by reducing the gain, which however reduced accuracy. Other methods of stabilisation include adding a passive phase lag network ($P + I$) or phase lead network ($P + D$) or a stabilising feedback *CR* circuit with the object of increasing stability without reducing gain. Curve d moved to curve e of Fig.

15.8 illustrates the objective. It should be noted that good stability and high accuracy are incompatible and a compromise between the two is desired for best performance.

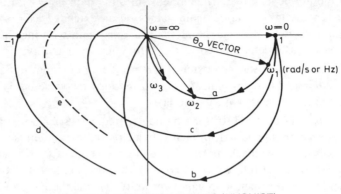

Fig. 15.8   RESPONSE CURVES (NYQUIST)

Contours of open loop response (Figs 15.7 and 15.8) can be used to evaluate closed loop response. For any point on the open loop, vector values of output to input ratio and the angle between them give points for the closed loop response. These can be plotted on another similar harmonic response diagram or on a frequency response diagram.

The Nichols chart (Fig. 15.9) is frequently used in frequency response and stability analysis. Point plots of constant gain and phase (derived by calculation due to the logarithmic scale) give contours on which open loop response can be plotted, minimum contours are shown on the sketch to simplify the illustration. Note that $20 \log_{10} 1 = 0$ dB, $20 \log_{10} 10 = 20$dB and $20 \log_{10} 0.1 = -20 \log_{10} 10 = -20$dB for the dB plotting so that negative dB corresponds to amplitude ratio (gain) of less than 1 (attenuation). Magnitude ratio 0·7 to 0·4 (−3 to −8 dB) corresponds to gain margin 0·3 to 0·6 (3 to 8 dB) which with a phase margin about 45° is a typical simple design specification.

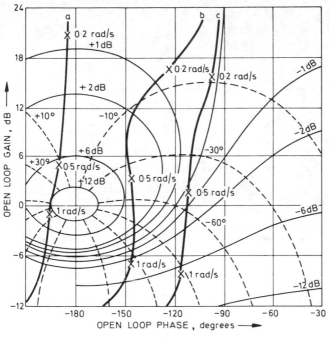

Fig. 15.9 NICHOLS CHART

On Fig. 15.9 curve a (unstable) shows 15° phase lead at 11 dB, curve b (stable) 32° phase lag at −3 dB and curve c (improved stability 63° phase lag at −8 dB, the operating frequency to be near 1 rad/s.

The closed loop characteristic (direct feedback systems) can be derived on this chart from the intercepts of the open loop locus with the gain and phase contours. This requires a simple calculation for values (which includes a feedback fraction).

FURTHER ANALOGUES

In some cases approaches are used in which electrical components in parallel are regarded as equivalent to mechanical components in series, and vice versa. Certainly current, which has a common value through series electrical components, is analogous to velocity, which has a common value through linked mechanical translational components. Capacitors in parallel are equivalent to springs in series, and vice versa.

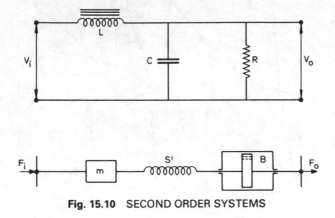

Fig. 15.10   SECOND ORDER SYSTEMS

The lower diagram of Fig. 15.10 illustrates a mechanical translational damper, spring, mass system which has been described previously (the left hand end is often fixed, $F_i = 0$, $F_0 = F$). It is typical of an anti-vibration mounting in which high frequency input oscillations will be damped out (see analogue circuit diagram, Fig. 16.1B)

The upper diagram of Fig. 15.10 is the equivalent electrical circuit. High impedance to high frequency inputs results in a filter system in which only low frequency components are applied to R. The ratio of V_0 to V_i can be found at a given frequency.

Fig. 15.11 illustrates the systems approach to a factory management system which should be self explanatory. Management includes finance, development, etc. sub-systems and techniques include critical path analysis, O.R., quality control, O & M, queuing theory, etc.

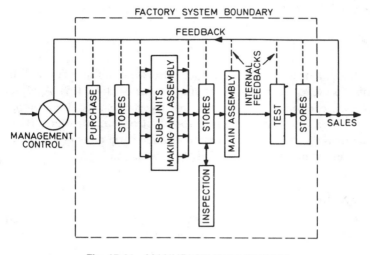

Fig. 15.11 MANUFACTURING SYSTEM

COMPONENT INTERACTION

By this heading is meant the internal interference within a controller of the interference effect between connected controllers.

INTERNAL INTERACTION

The three term equation for a controller has been considered and the equation in transfer function terms is:

$$V = -K_1\left(1 + \frac{1}{SD} + TD\right)\theta$$

This assumes each control action can be generated separately which may not be possible, and resulting interaction can occur which will affect output signal. Let consideration be applied to a three term pneumatic controller - see Fig. 10.7 of Chapter 10. As stated the position of integral and derivative adjustment can affect output in various ways. Consider the derivative adjustment placed at X on Fig. 10.8. The control output signal can be shown to be:

$$V = -K_1 \left[\left(1 + \frac{2T}{S} \right) + \frac{1}{SD} + TD \right] \theta$$

Integral or derivative adjustment affects controller gain (K_1). In this case K_1 is altered by the interaction factor $(1 + 2T/S)$. Analysis of each controller design is required to establish the exact output signal. Similar remarks apply for electronic controllers. Bode diagrams can be utilised to obtain characteristic plots of controller response which exhibit gain and phase variation.

EXTERNAL INTERACTION

Consider two first order controllers in series with *no* interaction between stages. The overall transfer function becomes:

$$\frac{\theta_0}{\theta_i} = \frac{1}{(1 + \tau_1 D)(1 + \tau_2 D)}$$

which is a non oscillatory type of second order function. For a sinusoidal input the gain (attenuation) of individual elements are multiplied and phase angles added algebraically, utilising as usual $i\omega$ for D.

The control units must be non interacting otherwise the transfer function of one controller will be modified by the loading of a following controller. This is usually avoided by inserting buffer amplifiers (or stages) of unit gain, without phase shift, between the controllers in series.

BLOCK DIAGRAM REDUCTION

A single (overall) transfer function can be obtained from a complete system consisting of individual (non-interacting) units each with its own transfer function. Block diagram algebra is used with two basic rules applied to each block pair in successive reduction i. parallel blocks are summed ii. series blocks are multiplied. A system and solution are detailed in Fig. 15.12:

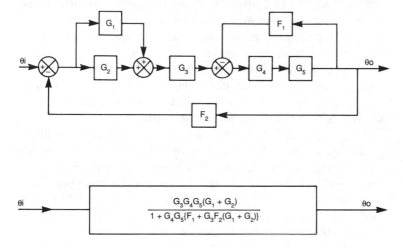

$$\frac{G_3 G_4 G_5 (G_1 + G_2)}{1 + G_4 G_5 \{F_1 + G_3 F_2 (G_1 + G_2)\}}$$

Fig. 15.12 BLOCK DIAGRAM REDUCTION

COMPONENT ADJUSTMENT

The adjustment of controllers, especially $P + I + D$, in a plant is usually done empirically by generally well established experience criteria.

ADJUSTING CONTROLLERS TO PLANT

For initial commissioning the controller must be set up exactly to the manufacturers instructions and all maintenance must follow similarly from makers advice. Integral resistance is usually set at maximum and derivative resistance at minimum. Proportional band can now be set for minimum stabilisation time. Derivative resistance can now be increased to reduce this time a little, integral resistance now being adjusted to the same as derivative resistance. There is a definite relation between T and S settings; even with independent settings T can never exceed S.

Considering now setting and adjustment in more detail. This is a skilled operation requiring time and a knowledge of plant characteristics so that the following, for a $P + I + D$ controller, is obviously a condensed simplification.

The object is to critically damp the signal to rest in the minimum time without overshoot and oscillation. Instability may occur for too narrow a proportional band, too short integral action time or too long derivative action time. Stability with underdamping gives oscillation with too long a stabilisation time; overdamping gives no oscillation but too long stabilisation time usually due to a too wide proportional band, too large integral action time or too short derivative action time.

For proportional action band only, it is best to start at say 200% bandwidth and move the dial away from and then back to the set value, noting the settling time. This is repeated at step reductions of bandwidth until the oscillations do not reduce to zero (too much reduction would cause instability with increasing oscillations). A slight increase in bandwidth now gives the correct value for minimum offset and stabilisation time.

For $P + I$ the proportional band would be set as for P action above with integral action time at maximum. The integral action time is then reduced (using big steps initially) until hunting oscillation starts. A slight increase in integral action time now gives the correct value and stability.

For $P + D$ the proportional band is narrowed as above until hunting is occurring and it is held at that value. The derivative action time, which had been set at minimum, is increased to remove hunting. The proportional band is again narrowed slightly and hunting removed by adding to the derivative action time. This process is continued until hunting cannot be removed by the derivative action time. The proportional band is now widened slightly for correct setting.

$P + I + D$ controllers may be adjusted in practice as for a P + D controller, noting the derivative action time, add the same integral action time then adjust for minimum offset. Interaction is always rather a problem except in a well designed controller, well matched to plant characteristics.

EMPIRICAL SETTING METHOD ($P + I + D$)

I and D terms are reduced to zero and proportional band is narrowed until continuous cycling occurs. This may require a small step input on the desired value setting to start the oscillation. Whilst continuous oscillation at constant amplitude is taking place the periodic time is measured. With this proportional band (W) and periodic time (T_0) it is possible to empirically set the controller thus:

$$\% \text{ Bandwidth for } P = 2W$$

$$\% \text{ Bandwidth for } (P + I) = 2.2W$$

$$S \text{ for } (P + I) = \frac{T_0}{1 \cdot 2} \text{ minutes}$$

$$\% \text{ Bandwidth for } (P + I + D) = 1 \cdot 67W$$

$$S \text{ for } (P + I + D) = \frac{T_0}{2} \text{ minutes}$$

$$T \text{ for } (P + I + D) = \frac{T_0}{8} \text{ minutes}$$

Signals are then trimmed for optimum performance. The above is satisfactory for a continuous process but not for auto-start with no overshoot.

TEST EXAMPLES 15

1. Sketch the harmonic response (Nyquist) diagram for frequency response tests on:
    (a) a stable system,
    (b) a critically stable system,
    (c) an unstable system.

2. A thermocouple at 10°C is placed in a fluid at a temperature of 60°C and the reading after 4s is 40°C. Assuming exponential delay response evaluate the time constant of this instrument. If the thermocouple were then used to measure a temperature rising steadily at 2°C per second what would be the steady-state error of the reading? (4·37s, 8·74°C)

3. A step change of 2·5% is applied to the input of a $P + I$ controller and the output gives a sudden step change of 5% and after two minutes the total output change is 12·5%. Determine proportional bandwidth and integral action time. A ramp change of 1% linearly is applied to the input of a $P + D$ controller and the output gives a sudden step change of 5% and after this the output changes linearly at 3% per minute. Determine proportional band-width and derivative action time. (50%, 80s, 33%, 100s)

4. Detail a "trial and error" (or "hunt") method of setting a three term controller. Show on a diagram the effect of i. too long ii. too short and iii. correct setting of integral action time for a $P + I$ controller.

CHAPTER 16

LOGIC AND COMPUTING

This chapter will be concerned with analogue computers, switching logic circuits, digital computers, data processing and computer control. Each section is obviously a specialist area of study and it is only possible in this chapter to give an introduction to each.

ANALOGUE COMPUTERS

Many of the principles involved have already been covered in the immediately preceding chapters. It is now proposed to summarise this information.

ANALOGUE

That which has correspondence or resemblance (analogous) to something else which may be otherwise entirely different in form. Analogies between resistance-damping, inductance-mass, capacitor-spring, etc. have been considered. This is extremely useful for simulation. Similar relationships exist for pneumatic, thermal and fluid systems.

ANALOGUE COMPUTER

Essentially a device to represent continuous measures of physical quantities in numerical form. Provides concurrent, fast, inherently graphical and reasonably accurate simulation and investigation of mathematical models of dynamic systems. In most cases electrical analogues are used with voltage signals representing system variables.

Basic elements

These have all been considered previously in Chapters 10 and 11 utilising both pneumatic and electrical transducer analogues of variables. Elements include the operational amplifier, feed-back, averaging, ratio, summer, scalar multiplier, inverter, proportional-derivative-integral devices, flow, power, root extraction components, etc. incorporated as required to form a simple analogue computer.

Control engineering

The application of basic elements to form control devices in process and kinetic systems is a direct application of small analogue computation. These aspects have been described previously.

Data processing

The large analogue computer is used, certainly at input interface, of data recording, display, scanning, alarm, etc. systems. This aspect will be considered later in this chapter.

System analysis

The use of the analogue computer to solve mathematical equations for system analysis is a valuable engineering tool. Variables can be varied readily and simulated results quickly obtained. Outputs may be conveniently recorded on pen chart devices, two axes plotters $(X \sim Y)$, cathode ray oscilloscopes and ultra violet recorders. Obviously the mathematical analysis is usually involved but the selection of a simple analysis of a basic equation should effectively illustrate principles which can be applied to complex systems.

Simple pendulum analogue simulation

The procedure is an example of the systems approach to engineering problems. The procedure could be analysed from objectives as follows:

a. Sketch the *conceptual* (physical) *model* – see Fig. 16.1A.

b. Evaluate the *mathematic model* equations, in this case:

$$m\frac{d\omega}{dt} = -mg\sin\theta - mB'\omega$$

1. $$\frac{d\theta}{dt} = \omega$$

hence $$\theta = \int_0^t \omega\, dt + \theta_0 \quad i.e. \quad \theta = \theta_0 \quad at \quad t = 0$$

2. $$\frac{d\omega}{dt} = -g\sin\theta - B'\omega$$

hence $$\omega = -g\int_0^t \sin\theta\, dt - B'\int_0^t \omega\, dt + \omega_0$$

c. Draw *block diagram,* starting with integrators, adding scalars, etc. as required.

d. Scale variables, draw *analogue programme,* draw *circuit diagram* (see Fig. 16.1A).

e. Analyse by test for interaction, evaluation, synthesis; refine-iteration.

On the block diagram the multipliers, summers, integrators, inverter will be noted. It should be remembered that there is a sign change on the operational amplifier. Interconnection of components is termed *patching*.

On the scaled analogue programme (patching) diagram, sometimes called flow diagram, the unity ratio of amplifier resistance is indicated by a 1. Voltages V_1, V_2, V_3 and V_4 are analogues proportional to θ, ω, θ_0 and ω_0 respectively. Potentiometers shown P. Adjustment of potentiometer tap ratio k and amplifier gain K gives required ratio scaling for gain (F and G in feedback terms).

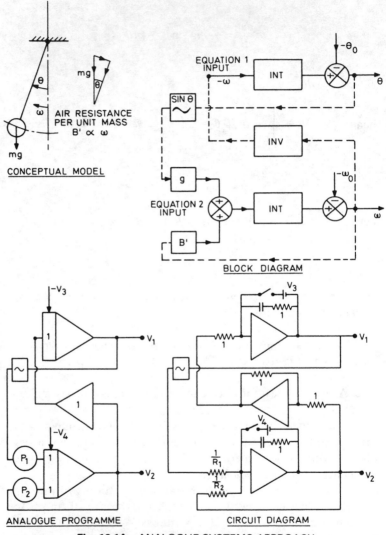

Fig. 16.1A   ANALOGUE SYSTEMS APPROACH

On the circuit diagram all factors should be clear from preceding diagrams. Resistances R_1 and R_2 are proportional to g and B'.

$$Dx = -ax + f(t)$$

$$x = -\int ax - f(t)dt$$

This is readily patched with two potentiometers (one input, one feedback) and an inverter integrator with suitable scaling for constants in a similar way to that shown in this pendulum example.

For a typical second order diagram:

$$(D^2 + B'D + g)\theta = 0 \qquad \text{pendulum}$$
$$(D^2 + 2c\omega_n D + \omega_n^2) x = a\omega_n^2 f(t) \qquad \text{general}$$

The approach is similar utilising, for example, three operational amplifiers in series (Fig. 16.1B). This example is an analogue for an anti-vibration mounting involving linear damper, spring, mass. The model diagram has been considered previously (see Fig. 15.10) and the equation, linear simple harmonic second order, was derived in Chapter 15 and is repeated in a different form (pendulum and general case).

Net force (mass times acceleration) acting in the opposite direction to displacement and velocity is given by:

$$m \frac{d^2x}{dt^2} = -S'x - Bv$$

(where S' is spring stiffness and B is damping coefficient)

$$\frac{d^2x}{dt^2} = -Cx - D\frac{dx}{dt}$$

$$x = -E\frac{dx}{dt} - F\frac{d^2x}{dt^2}$$

Referring to Fig. 16.1B:

The components used will decide values of constants. Scale factors have to be analysed relating voltage to variables – time constants allow real time scale variations. Output must equal input, so both are connected. An input force is necessary to provide essential acceleration and for this analogue is usually by square wave pulses from a signal generator. Output characteristics are analysed on an oscilloscope. Analogue variations (damping/mass and spring force/mass) are arranged at potentiometer and feedback resistor respectively. The circuit can be regarded from input to output as integration, or differentiation in the other direction. The last amplifier is a scale factor (which can also be arranged to have adjustable feedback).

It is now convenient to move from analogue (frequency domain) to digital (time domain) as used in logic circuits.

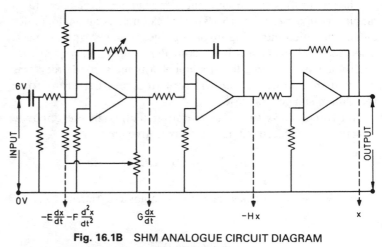

Fig. 16.1B   SHM ANALOGUE CIRCUIT DIAGRAM
(LINEAR VIBRATION DAMPER)

LOGIC CIRCUITS

Logic is a systematic approach to problem solving and decision making. Fig. 16.2 illustrates a simple logic flow chart analysis for marketing a new product (yes decisions full lines, no decisions dotted lines *i.e.* binary algorithm).

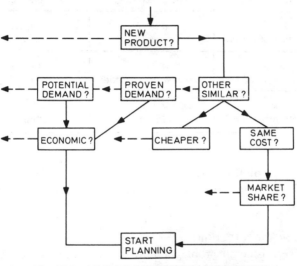

Fig. 16.2   LOGIC FLOW CHART (ALGORITHM)

BOOLEAN ALGEBRA

Is the algebra of logic. Has tended to the development of set theory, Venn diagrams, sentence logic and the logic of switching circuits. It is not possible to consider these aspects in any detail because notation and laws require involved consideration. However the table given (Table 16.1) indicates the obvious relationship and the logic of switching circuits can be developed somewhat to illustrate principles. Only two input variables (A and B) are considered for simplicity but extension is easily arranged.

Set Theory	Sentence Logic	Switching Circuits
$a \cap b$ Intersection of a and b	$a \wedge b$ a and b	$A \cdot B$ "and"
$a \cup b$ Union of a and b	$a \vee b$ Either a or b or both	$A + B$ "or"
a' Complement of a	$\sim a$ Not of a	$\overline{A}$ "not"

Table 16.1

The laws of Boolean Algebra and the use of Truth Tables greatly facilitate the simplification of electrical logic circuit design.

SWITCHING CIRCUIT LOGIC

Devices, using electromagnetic relays in control systems, have been used over a long period of time for such functions as sequential starting, protection interlocks, counting circuits, etc. A simple application of electrical circuits is shown in Fig. 16.3.

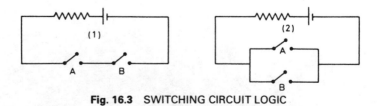

Fig. 16.3 SWITCHING CIRCUIT LOGIC

The truth table is given (Table 16.2) where "true" is referred to as state 1, *i.e.* relay closed, closed circuit, current flows, voltage across load; "false" is referred to as state 0, *i.e.* relay open, open circuit, no current flows, no voltage across the load. In electronics power supply connections are often +6V, –6V, 0V. Voltage used depends on devices and circuitry requirements.

A	B	A . B	A + B	$\overline{A . B} = \overline{A} + \overline{B}$	$\overline{A + B} = \overline{A} . \overline{B}$
0	0	0	0	1	1
1	0	0	1	1	0
0	1	0	1	1	0
1	1	1	1	0	0
INPUTS		AND	OR	NAND	NOR

Table 16.2

Note:

1. A series circuit is the AND function for output, *i.e.* output signal is the same sense as inputs when *all* inputs are the same only.

2. A parallel circuit is the OR function for output, *i.e.* output signal is the same sense as input change for any one or all input changes. *Inclusive* disjunction (gate) means either *or* all for the OR function and *exclusive* means either *not* all (symbol ⊕).

3. The single relay, or switch, is the NOT function, *i.e.* contact closed gives output (closed circuit, state 1, A) and contact open gives no output (open circuit, state 0, $\overline{A}$). Similarly the unity ratio operational amplifier (inverter) is a NOT function.

4. In electronics the provision of a unity ratio operational amplifier (inverter) in series with an AND circuit gives a NAND circuit (NOT-AND) with output opposite in sense to input when all inputs are the same only. Similarly the inverter in series with an OR circuit gives a NOR circuit (NOT-OR) with output opposite in sense to input change for any one or all (inclusive) input changes. Compare output states in the previous table. Logic

symbols vary, two conventions are shown in Fig. 16.4 and other forms are shown in the text to illustrate the variations. Small circles at output change AND to NAND and OR to NOR.

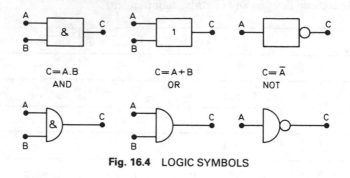

Fig. 16.4   LOGIC SYMBOLS

SOLID STATE LOGIC

Such logic circuits (gates) are being increasingly used in place of relays and thermionic valves. The convention for relays of 0 (off) and 1 (on) has to be modified because input and output have non clearly defined states less simple than on-off. 1 may represent higher (or more positive) voltage and 0 lower (or less positive) voltage. This is *positive* logic which is usually used with *npn* transistors because collector potential (and output) becomes more positive when the transistor is cut off. The reverse applies with *pnp* transistors and *negative* logic is used; **this convention is adapted in this chapter,** *i.e.* **negative true logic,** logical 1 negative with respect to logical 0 from a voltage level aspect. Digital logic functions can be achieved by the use of diodes and transistors, the former simpler and the latter more effective.

AND *gate (diode) negative true logic*
   Refer to Fig. 16.5:
   With A and B at say –6V (state 1) no current flows through R and output is –6V, *i.e.* output state 1 for coincidence of state 1s at input. With any input at say 0V (state 0), *i.e.* any diode conduct-

ing, output voltage is small, near 0 V (state 0 V). A diode, to earth across output, is sometimes fitted to ensure output 0 V. Reversing polarity of both input and output signal requirements gives an OR circuit, *i.e.* A or B or both at 0 V (state 0), with any diode conducting, means output is 0 V (state 0). Resistors are sometimes fitted at inputs (see Fig. 16.10).

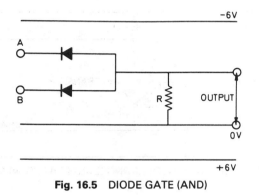

Fig. 16.5   DIODE GATE (AND)

OR *gate (diode) negative true logic*
   Refer to Fig. 16.6:
   Output volts are zero until one or both diodes conduct, when –6 V is applied to either or both inputs, the output is then –6 V, *i.e.* output state 1 for any combination of state 1 inputs.

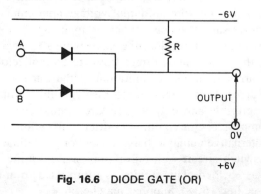

Fig. 16.6   DIODE GATE (OR)

Again reversing polarity reverses role (becomes AND) and resistors are sometimes fitted at inputs (with or without diodes – see Fig. 16.7). Diodes ensure that inputs cannot affect each other.

NOT *gate*
Refer to Fig. 16.7:

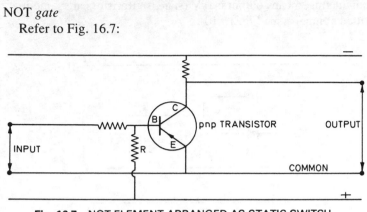

Fig. 16.7   NOT ELEMENT ARRANGED AS STATIC SWITCH

This is the inverter amplifier, single input, with output for no input and if any input no output. Input say –6 V (state 1) gives output 0 V (state 0), *i.e.* antiphase, or input 0 V output –6 V. For the configuration shown (*pnp* transistor) when there is 0 V input signal the base is slightly positive with respect to the emitter (reverse bias) and no current flows. The transistor is then essentially a high resistance impedance-resistance between emitter and collector with an output voltage approximately equal to the negative supply voltage. When an input signal of sufficient negative magnitude is applied the base swings to negative with respect to the emitter and current flows to the collector. If such current is arranged to cause saturation of the transistor then the resistance across emitter and collector is negligible so that the volt drop is negligible and output volts are almost zero. Thus the bistable amplifier with common emitter connection (as distinct from an alternative common base connection sometimes utilised) acts as a switch circuit with on-off limits. The emitter could be regarded as earthed. This device is often used in annunciator systems (see Fig. 16.18). Supplies may be ±6 V.

Referring to the **time delay** switch of Fig. 16.8 where operation is similar to the above. With no input volts, output volts approximately equal negative supply volts. When the input signal is applied and causes the base to become negative then output becomes zero as the transistor activates and the capacitor (CAP) charges. When the input signal is removed the capacitor discharges through the transistor emitter circuit as the rectifier (*A*) blocks any outlet (anode is negative) through the input circuit. So

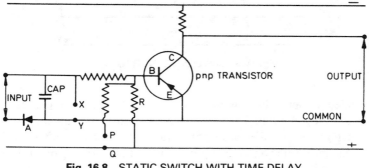

Fig. 16.8 STATIC SWITCH WITH TIME DELAY

there is a time delay before the output reappears, delay depends on the resistor and the capacitor values (*i.e.* time constant). A variable resistor between *P* and *Q* allows shorter time delays and a capacitor between *X* and *Y* allows longer time delays, resistor and capacitor both being adjustable. Supplies may be ±6 V for power source, common line may be earthed.

NOR *gate (inclusive)*
See Fig. 16.9:
The gate has one output and two or more inputs. Output changes if one, or more, input states change and the output change is antiphase potential to input change. This is a negative sign output OR gate. If –6 V is applied to *A* or *B* or both, the *pnp* transistor base is negative so that a large collector current flows and output volts are almost zero (small resistance across emitter-collector) if appropriate circuit values are designed in. With 0 V

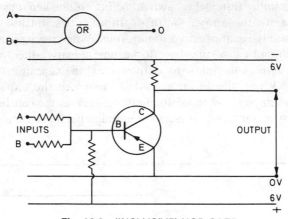

Fig. 16.9   (INCLUSIVE) NOR GATE

applied to inputs the output is –6 V (transistor biased to cut off at near +6 V). A symbol sketch is shown at the top of the diagram. Notation on such sketches often varies. An alternative often used is a semi-circle, base to inputs, with small circle (sometimes blacked in) or bar line on circumference leading to output to indicate inversion (Fig. 16.4).

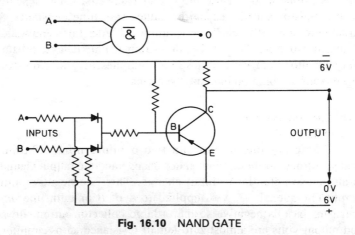

Fig. 16.10   NAND GATE

NAND *gate*

Shown in Fig. 16.10 with symbol sketch above.

Single output changes only if all inputs change. With inputs at 0 V output volts are –6 V and transistor is biased at cut off. If all inputs are –6 V then bias is removed and the output is 0 V. Output and input are antiphase. This is a negative sign output AND gate.

Note:

1. Circuit simplification will be apparent if only one (or two) gates are used for all duties. NAND or NOR gates are so used.

2. Reversing polarity of gates reverses the role of the device.

3. NAND gates can act as inverter NOT gates when only one input is applied. If a NAND gate output is fed to another single input NAND gate in series final output is in phase, *i.e.* AND gate. Combinations of NAND gates can also provide OR, NOR, HOLD, circuits.

4. Similar remarks to 3. apply in general for NOR gates.

HOLD *(memory) circuit*

A time delay circuit has been considered previously (Fig. 16.8). The remaining logic circuit commonly used is hold, or memory, which can now be considered. Feedback holds in the circuit even after input signal is removed. A reset signal then restores the state of the system.

Consider Fig. 16.11:

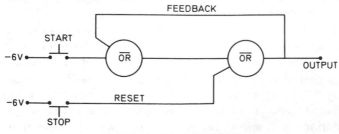

Fig. 16.11   HOLD CIRCUIT

Closing the start button momentarily allows –6 V input and the first NOR gate conducts with output 0 V. The second NOR gate gives output –6 V and this signal is fed back to the first NOR gate. This circuit output is maintained when the start button is released, *i.e.* held or remembered from the initial signal instruction. Closure of the reset (say stop) button changes the 0 V line input from reset to –6 V. The second NOR gate, previously with two inputs at 0 V, now has one input at –6 V and conducts, so output is 0 V. Final output becomes 0 V and feedback at 0 V means both inputs to the first NOR gate are 0 V and its –6 V output maintains the non-energised state when the stop button is released.

Another alternative example is given in Fig. 16.12 following.

Electronic transistor gates are similar to relays. Functions can also be achieved using diodes powered by the input signals. However transistor gates, as amplifiers, have the decided advantage of utilising a separate circuit of supply for output power which increases the scope greatly.

Note:

In logic circuitry entire circuits are packaged and it is not necessary to know the exact circuit configuration of a particular device (chip) because it is encapsulated. Signal tracing is impossible and it is only necessary to understand the relation between overall input and output signals and repair is by replacement (the black box philosophy).

NAND and NOR are obviously combinations of the three given actions and various "tree" type circuits can be quickly built up for otherwise complicated functions.

For example the logic illustrative circuit shown as a combination in Fig. 16.12. The circuit may be interpreted as follows: "if the *off* signal is not interrupted at the button and the *on* signal or the *feedback* (or interlock) signal G is energised there will be an output at H. Pressing the *on* button gives inputs at A and D, hence an AND function and output at H (the two NOT functions at B and C cancel, *i.e.* like two negatives make a positive; similarly E and F). Release of the *on* button still allows output to be maintained through the *feedback, i.e.* the alternative input of the two element (OR) circuit. Pressing the *off* button cuts

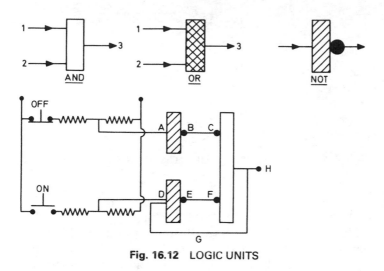

Fig. 16.12 LOGIC UNITS

off one of the signals in the AND circuit and cuts off output H." Strictly the combined sketch is NOT (A), NOR (D), NAND (CF) but redundant items ($BCEF$) simplify to OR (D) and AND (CF) and the shading section on D would then be crossed.

Flip-flop circuit

Multivibrator circuits have been discussed (Chapter 7). The univibrator circuit as sketched in symbolic form in Fig. 16.13 is used in computers. With inputs X and Y at state 1, say –6 V, the feedbacks at state 0 (0 V) and state 1 (–6 V). For the lower gate feedback (0) and input Y (1) through NAND gives state 1 (–6 V). For the upper gate feedback (1) and input X (1) through NAND gives state 0 (0 V). Connecting inputs to state 0 will give stable reversing. A positive pulse to one input reverses output potentials again. For counting both inputs are connected so that two positive input pulses are necessary for each positive output at the outer transistor, *i.e.* each successive binary changes at *half* the speed of the one driving it.

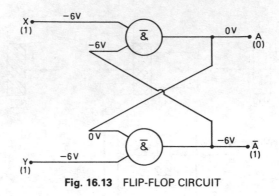

Fig. 16.13   FLIP-FLOP CIRCUIT

LOGIC SEQUENCE ENGINE SYSTEM

Figure 16.14 is a practical illustration of applied logic. All inputs, points 1 to 7, can be regarded as –6 V through input resistors. With all essential start criteria 2 to 6 satisfied the transducers will have closed these input relays and operation of the start button 1 gives all 0 V inputs (earth) to gate *A* whose inverted output is –6 V. Gate *B* inputs are –6 V from *A* and –6 V from the inverted feedback NOR gate (whose inputs are all 0 V). Gate *B* output at 0 V supplies the air start signal pulse and via

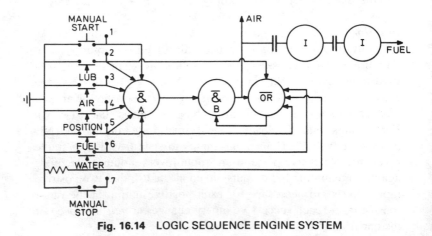

Fig. 16.14   LOGIC SEQUENCE ENGINE SYSTEM

double inverters *I* (single input NOR or NAND gates) gives a delayed fuel supply pulse (delay by capacitors in circuit) at 0 V. All inputs to NOR gate at 0 V. If any feedback input to NOR gate becomes –6 V (due to stop button operation or fault on circuits 2, 5, 6) its output becomes 0 V to gate *B*. Gate *B* will stop conducting and its output signal will become –6 V so shutting off fuel by signal change. A time delay, not shown, can easily be arranged to shut off only the air start signal after use. Many versions of the above circuit can obviously be built with various logic gates.

FLUIDIC LOGIC

Devices use fluid flow in sensing, logic computation and actuation. A commonly utilised principle is the Coanda effect, *i.e.* the tendency of a jet stream to attach itself to the pipe surface (wall). A complete range of devices can be built up and are widely used in industry. It is not possible to cover all designs but to illustrate principles two types are chosen. The bi-stable multivibrator (flip-flop) has been considered electronically and fluidic principles are illustrated in Fig. 16.15.

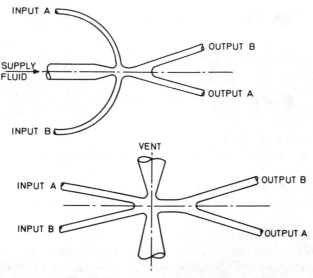

Fig. 16.15 FLUIDIC LOGIC DEVICES

Referring to the top sketch the fluid can be be regarded as attached to the wall of tube *A* at output. Input signal at *B* separates attachment and diverts flow to tube *B* output. This condition will hold (memory) if input *B* signal is removed and can be reset by an input *A* signal to the initial state.

The lower sketch illustrates an exclusive OR device, *i.e.* input *A* or *B* will give output *A* or *B* but not both. If inputs are zeros output is zero and if inputs are applied together the jets impinge, flow goes to vent, output to zero.

If output change for this momentum type device is greater than input (control) change it is amplification. A turbulence amplifier utilises a cross control jet to change from laminer flow.

DIGITAL COMPUTER

It is first necessary to consider simple computer-counting in binary terms.

SCALES OF NOTATION

The *denary* scale (base 10) is in general use but any number base can be used in a scale. Consider the following:

$$\text{Scale 10} \quad \ldots \quad 10^3 \quad 10^2 \quad 10^1 \quad 1 \quad \frac{1}{10} \quad \frac{1}{10^2} \quad \frac{1}{10^3} \ldots$$

$$\text{Scale 3} \quad \ldots \quad 3^3 \quad 3^2 \quad 3^1 \quad 1 \quad \frac{1}{3} \quad \frac{1}{3^2} \quad \frac{1}{3^3} \ldots$$

$$\text{Scale } n \quad \ldots \quad n^3 \quad n^2 \quad n \quad 1 \quad \frac{1}{n} \quad \frac{1}{n^2} \quad \frac{1}{n^3} \ldots$$

51 denary in the scale 3.

$\quad 51 = 1 \times 3^3 + 2 \times 3^2 + 2 \times 3^1 + 0 \times 1 = 1220 \text{ (base 3)}$

248 denary in the scale 5.

$\quad 248 = 1 \times 5^3 + 4 \times 5^2 + 4 \times 5^1 + 3 \times 1 = 1443 \text{ (base 5)}$

1475 denary in the scale 12.

$\quad 1475 = \text{ten} \times 12^2 + 2 \times 12 + \text{eleven} \times 1 = t2e \text{ (base 12)}$

(*t* and *e* to be new digits representing ten and eleven in the denary scale)

19 denary in the scale 2.

$19 = 1 \times 2^4 + 0 \times 2^3 + 0 \times 2^2 + 1 \times 2^1 + 1 \times 1$

$= 10011$ (base 2)

The latter, base 2, is the *binary* scale.

BINARY SCALE

The reader should first verify for practice:

101011 base 2 = 43 base 10

39 base 10 = 100111 base 2

This scale has the advantage of expressing any number in terms of two symbols, 0 and 1. If 0 is taken to represent current off and 1 to represent current on (or two voltage states) then it is possible to use many electrical currents to perform calculations. This is the basis of the **digital computer** counting in digits, pulses or bits.

Addition and subtraction

htu		2^6	2^5	2^4	2^3	2^2	2^1	1
43			1	0	1	0	1	1
+ 39		+	1	0	0	1	1	1
1	Carry Figures	1			1	1	1	1
82		1	0	1	0	0	1	0

Addition is as illustrated above. If the digit sum in any column totals 2, carry 1 to the next column and leave 0; if 3 carry 1 leave 1; if 4 carry 1 two columns and leave 0; etc.

Subtraction is as illustrated below. When subtracting 1 from 0 borrow (10) *i.e.* 2 from next column, change 0 to 1 working to the left until a 1 is reached, change this to a 0, and continue the subtraction.

htu		2^6	2^5	2^4	2^3	2^2	2^1	1
Borrow Figures			1	(10)	(10)	1	1	(10)
88		1̶	0̶	1̶	1̶	0̶	0̶	0
−63	−		1	1	1	1	1	1
25				1	1	0	0	1

Multiplication and division
Procedure exactly as denary using final addition or subtraction techniques above.

```
      1 0 1 1                    1 0 1 0
    × 1 0 1 0          1 0 1 1 | 1 1 0 1 1 1 0
    ─────────                    1 0 1 1
      0 0 0 0                    ─────
    1 0 1 1 0                        1 0 1 1
  0 0 0 0 0 0                        1 0 1 1
1 0 1 1 0 0 0                        ─────
─────────────
1 1 0 1 1 1 0
```

COMPONENT UNITS

A brief introduction to constituent elements of the digital computer can first be considered. Consider Fig. 16.16:

Input unit

Functions to accept input data coded information in the form of either punched or magnetic tape or converted analogue signals and transmit the electronic signals in digital form to the arithmetic unit. The connecting link between units is called a *bus*. *Programs* are typed in on a keyboard and displayed on a *video display unit* (VDU).

Arithmetic unit

Processes inputs to desired arithmetic functions such as addition, averaging, etc. Assembly of bistable devices for binary operation including binary counter (flip-flops in series, extending range), shift register (flip-flops driven by clock), accumulator and buffer store, etc. all pulse operated combinational and memory devices. Grouped with memory and control units to form the *central processing unit* (CPU) which if a single chip would be the microprocessor in a microcomputer.

Memory unit

Store information with basic two stable state element. Storage can be paper or magnetic tape spools, magnetic drum or magnetic core stores. Magnetic cores are of two state ferrite core rings, grouped into lines (*words* or *bytes*) with up to 36 digits (*bits*) per word. A store with 12 bit (cores) per word and 2^{12} words (4096) has almost 50 k cores and is known as a 4 k memory store. This would be classified as small with 64 k as quite large. *Random access memory* (RAM) is where instructions in or read out. ROM is read only memory.

Control unit

Essentially the brain element of the computer. This is sometimes referred to as programme unit. The programme input is decoded, addressed and so internal transfer involving memory and arithmetic functions is carried out. Programmes involved in the central processor may be specific *language* for a particular type of machine or generalised high level language code such as Algol,

Cobol, Fortran, which is processed to machine code. Input and output units are also controlled. Assemblers and compilers translate between languages.

Output unit

Receives computed outputs as electrical signals, transmits directly to control functions of plant, operates digital logging units, display and alarm devices and presents output for typewriter or as tape. Outputs may require reconversion digital to analogue for control action to analogue computers, plant controllers, etc.

Peripheral units

In this case is meant outside the digital computer itself *i.e.* outside the input-output boundaries shown chain dotted on Fig. 16.16.

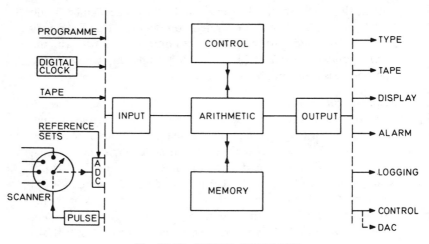

Fig. 16.16   DIGITAL COMPUTER

For general purpose computers requiring high speeds and used mainly for data processing in business systems inputs and programme inputs would be tape and outputs tape and type. The programme input and digital pulse timing clock are obviously essential peripheral equipment to any digital computer.

Engineering computers require reference sets, scanners and

analogue-digital converters. Outputs would be type logging on paper tape, display, alarm and control functions.

Reference sets

Amplified transducer inputs from scanners are compared to manually set high-low level limits by potentiometers arranged as multi-position rotary switches before analogue inputs to the converter. Alternatively comparison is arranged digitally by direct programme control on desired values or manual set by pins inserted on a matrix patch board.

Scanner

Measurements are sequentially selected by transistor circuits. The transistors are operated by an input signal from the scan control device regulated by pulses from a pulse unit digital clock. Scan speeds can reach 400 points per second if required.

Analogue to digital converter

Analogue display is illustrated by a car speedometer, *i.e.* continuous, and digital display by the distance device, *i.e.* discrete steps. One design of converter has a reference voltage potentially divided by resistance binary steps 2^0, 2^1, 2^2, etc. which are compared to the input voltage signal in sequence, until parity, when output from tapped resistances is then digital. A similar principle can be applied by a balance bridge comparing input and output resistances. There are many types of converter available and a typical digital instrument is shown in Fig. 16.17. The comparator provides an output when the reference voltage is greater than the integrated input voltage. This output opens a gate

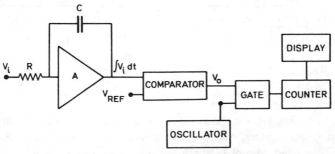

Fig. 16.17   DIGITAL INSTRUMENT

circuit allowing the oscillator pulses (proportional to input voltage) to the counter. Range variable, with RC time constant

DIGITAL COMPUTER

Units and sketch have been considered. In large computers it is necessary to have maximum utilisation of computer time. Batch processing utilises storage until a suitable time for handling the programme. Real time operation requires instant response to priority input when required. Time sharing to consumers is generally necessary in present digital systems.

DATA PROCESSING

Main application is in business systems. In engineering, digital data processing equipment provides an integrated system of monitoring plant process and includes alarm scanning, centralised display and data recording.

DATA LOGGER

This is essentially centralised instrumentation only. As such has now only a secondary function of measuring and logging with display. Useful for continuous monitoring and documentation of records of, for example, large refrigeration storage plant. Generally refined and superseded by an integrated system to include alarm scanning which provides malfunction protection in complex machinery installations.

ANNUNCIATOR SYSTEMS

Are central surveillance – collection points. An off normal condition results typically in the following sequence

Condition	Green Lamp	Red Lamp	Klaxon
Normal	On	Off	Off
Fault	Off	Flashing	On
Receive	Off	Steady	Off
Normal	On	Off	Off

Fig. 16.18 shows a diagrammatic circuit for a single point annunciator system. State (1) is –6 V and State (0) is 0V, NOR gates are used (except J–NOT).

Consider Fig. 16.18:

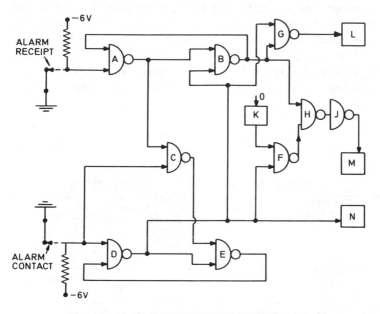

Fig. 16.18 ANNUNCIATOR SYSTEM

Under normal conditions the alarm contact is closed, D output (1) which operates the green light N, G output (0) and J output (O) so klaxon L and red light M are off.

At fault, alarm contact opens, D output (0) and N off, E output (1) which feeds back to hold D. Input to F is (0) and signal generator K also enters pulses into F so that M flashes. Outputs from D and B are both (0) so output from G is (1) and operates L.

When the alarm receipt button is operated a (1) signal is fed into A and B, output is (1) which feeds back to hold A; also overrides flash input to H so M exhibits constant red light and switches L off via G.

When the alarm contact returns to normal and closes, both inputs to C are (0) and its (1) output resets the hold circuit DE, then AB, so M is extinguished and N lights up.

INTEGRATED SYSTEM

Includes data logging as one function. Essentially a digital computer but of a fairly simple design, peripherals are as already described. The programme is designed in and fixed so that programme input, control, arithmetic and memory units as shown in Fig. 16.19 are simplified and combined into one central processor. A very brief outline description will now be given and reference to Fig. 16.16 and Fig. 16.19 should be made.

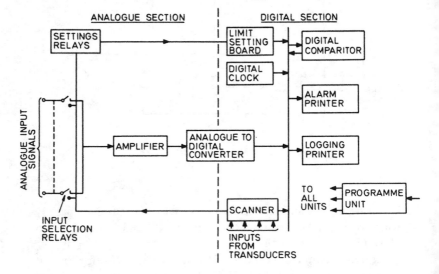

Fig. 16.19 SCHEMATIC DIAGRAM OF ELECTRONIC ALARM SCANNING AND DATA LOGGING SYSTEM

Primary inputs

Analogue signals from transducers represent variables such as pressure, level, flow, etc. Each transducer has a plug-in printed circuit module for measurement range, limits, etc. and is switched in by the scan control unit.

Signal selection

A scanner unit. Sub-units may be used at local points in the installation for say 40 points which reduces cable runs to the central control room and relieves space requirement of the processor.

Signal processor

Provides amplification, analogue to digital conversion, limit sets analogue or digital, scaling, linearising, etc. with outputs to digital (or analogue with d-a conversion) display, routine logging and alarm circuits.

Output devices

Visual and audible alarms are required. A logging typewriter records alarm conditions and events, until cleared. The typewriter can log by routine all points on demand or at time intervals. *Display* requires careful consideration with a minimum of gauges restricted to essential observation. Mimic diagrams of plant circuits with indicator lights to indicate key points in the plant are often used. A central control room needs careful design for comfort, correct instrumentation and alarm indication plus good effective lighting. Construction of components is modular with plug-in circuit cards for fault rectification. Process plant subject to such surveillance provides accurate and regular records and machinery protection with reduced watchkeeping staff.

COMPUTER CONTROL

This envisages a full digital computer so programmed to maintain output variables at the design condition by providing outputs to controllers. At present most controllers work on analogue inputs so the digital computer requires an output digital-analogue converter. Direct digital control is however coming into use where computer output acts directly as control action on the final correcting element. Simple computers generally include controllers, data processing is now well established, it follows that computer control is very relevant. In marine practice a computer can be so programmed to provide for example complete preparation of machinery plant for sea together with computer control on passage. Built in emergency action is provided. Such a large computer could of course cope adequately with cargo handling in tankers and navigational route placing with suitable addition to store and provision of correct programming. The analogue computer is faster on a complex path run but the digital computer is accurate, repeatable and suited to short runs. There are many fairly cheap analogue-hybrid units and parallel hybrid refers to the addition of digital logic for automatic programming of variables. Interface matching between analogue and digital can often be very difficult. A true hybrid computer is digital: interface: analogue which usually requires a suitable compiler processor language for operation such as Actran.

Recent developments have led to increased automated mass production of miniaturised and highly sophisticated functional complexes of circuits at a decreasing cost and volume (size). Mini computers of large potential computing power are now used with time sharing links to central large main frame computers. Micro processors, with a memory chip (storage, programme) and input/output devices linked to the central processing unit chip, are increasingly used virtually as micro computers dedicated to one particular control system and function.

COMPUTER SIMULATION

Overall system performance, including interaction between components and loops, at the initial design stage is becoming increasingly important. Analysis and simulation of the dynamic (transient) performance, as well as steady state behaviour, is required. Mathematical models, based initially on linear analysis for frequency response, and by computer simulation for non-linear systems are applied in design studies.

The upper sketch of Fig. 16.20 illustrates a boiler water-level control loop. Boiler (B), steam (S) and feed (F) flow transmitter signals are to the P computing relay (X) with output, joined by level transmitter signal (L), arranged to the $P + I$ computing relay (Y). The output signal from Y operates the feedwater control valve and positioner (V).

A computer simulation can be set up and a suitable patch diagram is shown in the lower sketch of Fig. 16.20.

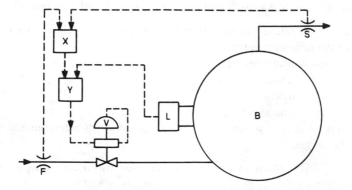
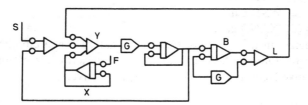

Fig. 16.20   BOILER WATER LEVEL CONTROL AND SIMULATION

TEST EXAMPLES 16

1. Describe, with sketches, an electrical panel for use in monitoring the alarm conditions of a set of diesel generators.

2. With references to data logging systems, explain the meaning of the following terms:
   (a) sensing device,
   (b) scanner,
   (c) transducer,
   (d) scaling unit.
Make a diagrammatic sketch of the components interconnected in the logger.

3. Write down in logic symbols, for input and output, the following functions, each with two inputs A and B:
   (a)   OR gate,
   (b) NOR gate,
   (c) AND gate,
   (d) NAND gate.
Sketch, in logic symbol form, each gate. Draw a circuit diagram for the OR and AND gates and briefly describe the modifications to give NOR and NAND functions respectively.

4. For a data logger system:
   (a) what is meant by time division multiplexing?
   (b) explain the use of a comparator,
   (c) what is the difference between analogue and digital signals?
   (d) discuss the reasons for A to D conversion.

SPECIMEN EXAMINATION QUESTIONS

Note: Questions on instrumentation and control systems are set within a wide range of subject titles such as Engineering Knowledge (General, Motor, Steam), Instrumentation, Electrotechnology, Power Plant, etc. The specimen questions following are based on expected ½ hour answers, which is the most common practice. In some examinations however only short answer questions are set, e.g. Class Three (DTp – SCOTVEC) – 10 minutes, and in others a combination of sections involving ½ hour and short answer questions is used, e.g. some BTEC & SCOTVEC.

CLASS THREE (DTp – SCOTVEC)

1. Sketch a thermometer suitable for remote reading – indication.

2. Sketch a valve for automatically controlling fluid flow in a pipeline.

3. Describe, with a simple sketch, any level indicating device.

4. An engine alarm sounds intermittently. How would you determine if it is a genuine or nuisance alarm?

5. List any five essential instrument readings, indicating the running condition of machinery, for display on a centralised instrument panel.

6. Describe briefly any two types of pressure measuring device.

7. Sketch a moving coil type of electrical measuring instrument.

8. Briefly describe any method of speed control for an electric motor.

9. Describe, with a simple sketch, how a flapper-nozzle device controls pneumatic pressure in an air line.

10. Describe how remote smoke/fire indication can be located at a central observation station.

11. List four essential parameters of machinery in operation requiring alarm indicators.

12. Describe, with a sketch, any one type of flame or fire or smoke detector.

13. Explain the principle of operation of a pneumatic diaphragm actuator.

14. Describe a device that would automatically activate an alarm if oil feed supply fails.

15. Sketch a pneumatic booster relay and explain briefly how it works.

16. With reference to automatic control technology, define and clarify the following:
      (a) Desired value;
      (b) Deviation;
      (c) Correcting unit.

SPECIMEN EXAMINATION QUESTIONS

CLASS TWO (DTp – SCOTVEC)

1. Describe with sketches, two methods for increasing the strength of a signal in a control system.

2. Describe instruments used for measuring the temperature in the following spaces:
   - (a) refrigerated hold,
   - (b) engine room,
   - (c) boiler uptake.

3. Describe with sketches, two methods for remotely determining the quantity of liquid in a tank. Compare the accuracy of these methods and explain how the degree of accuracy can be maintained. State one possible source of error for each of the methods described.

4. Sketch and describe three methods of sensing temperature change.
Describe how the signals are converted and fed into an automatic control system.

5. Sketch a compressed air system for use with pneumatic controls.
Write notes on each component shown.

6. Explain how a temperature sensitive element installed in a refrigerated stores locker can be used for starting and stopping a domestic refrigeration compressor.

7. Describe, with sketches, how the inlet temperature of lubricating oil to main engine or gearing may be automatically monitored and controlled.

8. Describe a bridge/engine room telegraph interconnecting gear. Explain how the system may operate a "wrong way" alarm.

9. State why a pneumatic control system requires clean dry air. Explain how the following air pollutants are dealt with:
    (a) water,
    (b) oil,
    (c) dust and dirt.

10. Draw a line diagram of an arrangement whereby the pressure of oil delivered to a main lubricating oil system by a constant speed, positive displacement pump is pneumatically controlled within set limits. Trace the sequence of events upon deviation in oil pressure.

11. Sketch a compressed air system for pneumatic controls labelling all the principal items.
Describe with sketches an automatic drain on the air compressor. State what routine maintenance and tests are needed to keep the system fully operational.

12. Sketch a pneumatically operated valve for regulating coolant flow.
Explain how the pneumatic system controls valve movement.
State how valve position is indicated at the control station.

13. Fig. A illustrates how a transistor amplifies a signal from a thermocouple to operate a relay. Describe the principles underlying this amplification and explain why it is necessary.

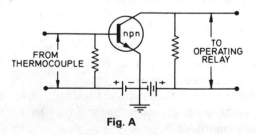

Fig. A

14. The following terms may be used to describe a boiler feed water controller:
   (a) detecting element,
   (b) servo-motor,
   (c) desired value,
   (d) difference element.
Relate the terms to practical components and describe their operation in the controller.

15. Sketch a simplified circuit diagram of an *npn* transistor illustrating its use as an amplifier. Give a reasoned account of its operation in terms of the electron theory.

16. Sketch and describe a master controller, operated by variation of pressure in the boiler, for regulating the air and fuel supply to the furnace by a pneumatic control system.
Explain how "hunting" of the system is prevented.

17. The left hand side of a small bar of crystalline silicon contains a small proportion of the element phosphorus as an impurity and the right hand side contains a small proportion of aluminium. Silicon has four valence electrons, phosphorus five and aluminium three.
Sketch the bar, and indicate the *n*-type region and the *p*-type region. Draw a battery connected between the opposite ends of the bar and show the polarity of its terminals which will bias the *p-n* junction in the forward direction.
Explain why no current will flow through the bar when this polarity is reversed.

18. With reference to medium or slow speed engines, describe transducers suitable for producing electrical or pneumatic signals to indicate:
   (a) lubricating oil pressure,
   (b) jacket cooling water temperature.
Explain how each of these transducers can be tested.

19. Fig. B illustrates a three phase full wave bridge rectifier circuit. Using this diagram explain how an ac voltage is converted to a dc voltage. Draw the graph of output voltage against time for one cycle. State where such a circuit could be used in marine engineering.

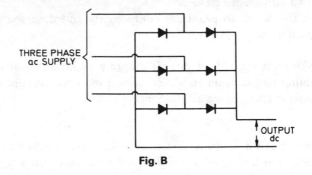

Fig. B

20. With reference to automatic combustion control, explain with sketches how:

(a) and why pressure differentials across air registers are measured,

(b) air/fuel ratios are adjusted,

(c) the boiler can continue to operate upon failure of the fuel flow regulating valve.

21. Sketch in detail a section through a hydraulic governor as fitted to medium speed unidirectional engines. Explain how it operates under frequent and wide fluctuations of load.

22. Describe with sketches an automatic supervisory and alarm arrangement for the control of bilge water accumulation under periodically unattended conditions.
Explain how the control system is prevented from giving a false alarm due to ship motion.

23. Explain the term "semi-conductor" and describe what is meant by "p" and "n" type materials.
With the aid of a simple sketch discuss the conditions at the junction between a p-type and an n-type semiconductor.

24. Draw a line diagram of a pneumatic system for control of ballast, labelling the principal components.

Describe how a ballast valve is remotely operated.

State why driers, filters and automatic compressors are preferable to "bleed off" from main or auxiliary air receivers.

25. Fig. C shows a simplified circuit of an automatic voltage regulator. Explain how the arrangement maintains a constant voltage output.

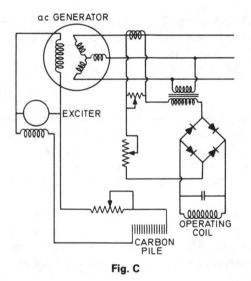

Fig. C

26. Describe, with sketches, integrated alarm and control systems for monitoring the following:
    (a) air pressure,
    (b) tank contents,
    (c) oil temperature.

27. Describe with sketches, two different methods for the remote measurement of fluid flow through a pipe. Compare the accuracy of the methods. Give one cause of error in each method.

28. Draw a line diagram of a system to operate widely distributed valves from one point.
Describe how any one valve is remotely manipulated.
Suggest, with reasons, a shipboard system which lends itself to such control.

29. Describe how a crankcase oil mist detector operates.
Sketch a mist detector system showing the run of the sampling pipes.
State how sampling is controlled.

30. Draw a line diagram of an automatic sootblower system.
Describe the sequential operation of the blowers. Give reasons for the order of operation followed.
State what safety devices are incorporated in the system and why they are fitted.

31. A bridge circuit containing zener diodes may be used in a static automatic voltage regulator to monitor the voltage output of an alternator. State:
    (a) details of operation of zener diodes;
    (b) how the error signal is derived in the bridge circuit.

32. (a) In practice "two step" control devices have a small "dead zone"; illustrate this by means of a control variable - time diagram.
    (b) Explain what is meant by "two step" control and give THREE examples of where such a control system may be used in an engine room.

SPECIMEN EXAMINATION QUESTIONS

CLASS ONE (DTp – SCOTVEC)

1. Describe how the supervisory equipment for the control of machinery in a periodically unattended engine room is itself monitored for defects on individual channels and as a complete unit.

2. Explain the problems in controlling the upper and lower limits of temperature of lubricating oil supplied to the main machinery.
Describe with sketches how this supply temperature can be automatically monitored and controlled.

3. State two advantages and two limitations in the use of electrical signals and also in the use of pneumatic signals as transmitting media in data transmission systems.
Explain with the aid of diagrams the principle of operation of a force balance transmitter employing either electrical or pneumatic signals as the transmitting medium.

4. Describe, with sketches, how electrical signals are converted to pneumatic signals in control systems.
Suggest a shipboard application featuring this conversion and state the defects to which the arrangement is susceptible.

5. Sketch and describe a system for indicating remotely the propeller shaft speed.
Explain how, for the system selected, inaccuracies occur and are kept to a minimum

6. Sketch and describe a method of measuring the pressure differential for fluid flow systems.
State what are the effects of altering the orifice plate size or the position of the tapping points.

7. Describe the construction and principle of operation of a Bourdon pressure gauge.

State the factors upon which its operation depends.

Outline a sequence of tests and adjustments applied to such a gauge known to be inaccurate and in particular mention at least three of the following:

    (a) leakage,

    (b) hysteresis,

    (c) non-linearity,

    (d) magnification,

    (e) zero error.

8. State what is the purpose of each of the following items in a machinery control system:

    (a) portable mercury manometer,

    (b) portable inclined-tube manometer.

    (c) portable temperature potentiometer,

    (d) compressor and vacuum pump.

Describe in detail any two of these items.

9. With reference to a gyro-controlled hydraulic steering gear, explain its action using control engineering terms making specific reference to:

    (a) desired value,

    (b) feedback,

    (c) actuator.

State the part played by the cylinder relief valves in the automatic control system.

10. Sketch and describe a fuel meter used with high viscosity fuel. Explain how it operates.

Explain the value of the readings obtained and how they are used.

11. Sketch and describe how coolant flow may be measured on a linear scale.

Explain the principle of operation of the instrument concerned.

Explain why the values recorded may vary from those expected from calculations.

12. State why the temperature of lubricating oil supplied to an engine needs close control.

Sketch and describe an arrangement and explain the principle of operation of instrumentation and control equipment for automatically maintaining the temperature of lubricating oil supplied to an engine at its desired value.

13. Sketch and describe in detail the construction and operation of one of the following:
  (a) an electric torsionmeter,
  (b) a preferential trip,
  (c) an electric telegraph.

14. Define the term "cascade control" as applied to control engineering.

Describe, with sketches, cascade control as applied to an engine coolant system. Show on the sketches how pressure and temperature varies at the cardinal points in the system.

Give one advantage and one disadvantage of this control arrangement.

15. Fig. D shows a circuit of a common-emitter amplifier. The transistor has a high current gain so that its base current is small. If the current through the emitter resistor is 0·4 mA determine the battery voltage. Assume that when the transistor is conducting, the voltage between the base and emitter is 0·2 V. (90·2 volts – this is abnormally high, with a 2 kΩ resistor in the emitter circuit the supply voltage is 11 V which is a much more acceptable value).

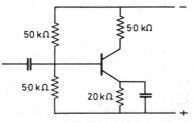

Fig. D

16. With reference to boiler combustion control explain:

(a) the operation of the master controller following variation in steam pressure,

(b) the importance of pressure drop across the air registers,

(c) how the air-fuel ratio is adjusted,

(d) how the boiler operation can continue upon failure of the fuel flow regulating valve.

17. The common-emitter output characteristics for a transistor are as follows:

Collector	Collector Current (mA)		
Voltage (V)	$I_b = 30\mu A$	$I_b = 60\mu A$	$I_b = 90\mu A$
3·0	1·0	2·1	3·2
7·0	1·29	2·55	3·9
10·0	1·5	2·9	4·4

Draw the graph of collector current against collector voltage and construct load lines to show the operation from a 6·5 V battery with load resistors of 1000 Ω and 1500 Ω respectively. If a suitable value of base bias current is 60 μA for an input for an input signal of ± 30 μA, determine the current amplification for each load. (33·3, 31·6).

18. The figure shows the equivalent T-circuit of a transistor used in a common-base circuit. The resistances presented to the alternating components of the current by the emitter (r_e), the base (r_b) and the collector (r_c) are 30 Ω, 0·6 kΩ, and 1 MΩ respectively. If the current amplification factor (α) is 0·98 and the load resistance (R) is 9 kΩ, calculate the current, voltage and power gain and the input resistance. (0·97, 186, 180; 48 ohms).

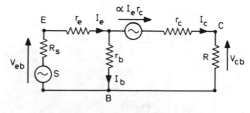

Fig. E

19. Describe with sketches how the pressure of a fluid is controlled by a pneumatic controller incorporating proportional and integral (reset) action.
Give reasons for instability in the controller action. State how this instability is overcome.

20. Give a detailed diagrammatic sketch of a mechanical-hydraulic governor. Explain how this governor operates. State what advantages it possesses over inertia governors.

21. Draw in detail a diaphragm operated control valve. Analyse the action of the inter-connecting elements, that is, the parts affecting control.
Explain how the load change is communicated to the actuator.
State where such a valve may be used in an engine room.

22. Explain why the simple float control feed regulator is inadequate for the present generation of main boilers.
Describe with sketches, or block diagrams, a feed control system in which it is possible to programme a set point for various loads.

23. Give the advantages and disadvantages of data-logging systems used in connection with ship's machinery.
Explain the value of recorded data and how it is interpreted and usefully employed.

24. With reference to automatic voltage regulators discuss the function of the following basic elements:
    (a) error detecting element,
    (b) correcting element,
    (c) stabilising element.

25. A pressure controller is fitted in a fuel line where ultimate state error is minimal and fast response is necessary. Sketch and describe such a controller making reference to components giving the desired characteristics.

26. Fig. F shows a simplified circuit of a transistor employed as a switch in a relay operation. Describe how this is accomplished.

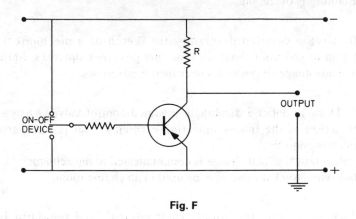

Fig. F

27. For the purposes of remote recording give two advantages and two disadvantages of:

(a) electrical signals,

(b) pneumatic signals.

Explain which system is best suited for sensing and remote indication of:

i. lubricating oil pump discharge pressure,

ii. engine bearing temperature,

iii. fuel tank liquid level.

State what precautions are necessary when sensors are located in hazardous areas.

28. Sketch and describe a three element feed water control giving reasons for its location.

Explain how unity relationship is maintained between the identified variables.

Explain why three element control is superior to two element control.

29. Draw a line diagram of a boiler combustion control system labelling the principal items.

Explain how the system functions and in particular how feed water supply, fuel supply and air/fuel ratio are regulated to match steam pressure and flow variation.

Explain how these controls can be tested for alarm conditions without upsetting the balance of the system.

30. Sketch and describe a hydraulic servo system associated with a controllable pitch propeller.

State how loss of fluid pressure occurs and warning is given of impending failure. Explain what happens upon loss of oil pressure and why.

State what routine maintenance is necessary to ensure trouble free operation of the propeller.

31. Explain how centralised control is achieved for either the direct reversing slow speed diesel engine or a main boiler and turbine installation. A block diagram could be used to show the principle functions and inbuilt safety features.

32. (a) With reference to primary sensing devices used in conjunction with a data logging system:

      i. describe the operation of TWO devices that are suitable for temperature measurement;

     ii. describe, with the aid of a sketch, a device that is suitable for pressure measurement;

    iii. describe, utilising a sketch, a device that is suitable for tank level measurement.

(b) Show, by means of a simple block diagram, the principle of data logging, and state why an Analogue/Digital convertor may be incorporated.

SPECIMEN EXAMINATION QUESTIONS

ONC – OND (BTEC & SCOTVEC)

1. Draw a labelled diagram of a flow measuring device to give a remote reading on an instrument panel.
Explain how the device operates to measure and indicate a change of flow.

2. (a) For any expansion type of thermometer, give four factors involved in obtaining a high speed of response.

   (b) List three methods of temperature measurement and sketch and describe one method in detail, giving advantages and disadvantages.

3. Describe, with a sketch, the static characteristics of a junction diode.
Sketch the forward and reverse characteristics and write short notes on:

      i. the forward resistance,
      ii. the leakage current,
      iii. zener action.

With the aid of waveform diagrams, illustrate the action of a capacitor connected in parallel with the output of a half-wave rectifier.

4. (a) List four different methods of measuring liquid level.

   (b) Sketch and describe a remote reading liquid level indicator suitable for a high pressure boiler.

5. (a) Draw a fully labelled sketch illustrating the principle of operation of a differential pressure transmitter.
Describe its operation for a change in the measured differential pressure.

   (b) State how zero and range adjustments are made on dp transmitters and name two different marine applications for such a transmitter.

6. Describe, with the aid of a suitable circuit diagram, a method of obtaining the output characteristics of a junction transistor.

Sketch the characteristics and from them show how it is possible to draw the transfer characteristic and hence determine the current amplification factor of the transistor.

7. (a) Draw diagrams of instruments capable of measuring:
      i. carbon dioxide content in a gas sample,
      ii. oxygen content in a gas sample.

  (b) Describe in each case, for these two instruments, the principle of operation and indicate typical readings expected in a sample of gas from a boiler uptake.

8. (a) State three essential requirements for any instrument suitable for use in a modern marine power plant.

  (b) Explain, with reference to instrument display, the terms "analogue" and "digital".

  (c) Give two advantages of the oscilloscope over moving iron instruments for the measurement of ac quantities.

9. (a) Draw the circuit diagram, indicating clearly the correct polarity of the electrical supplies, for a *pnp* junction transistor connected in common-emitter mode.

  (b) Draw the circuit diagram, indicating clearly the direction of current through the load, for four silicon diodes connected in a bridge configuration to produce full wave rectification of current in a resistive load.

10. Sketch and describe a pneumatic diaphragm operated control valve, clearly label and describe each part.

Explain what is meant by "Fail Open" and "Fail Closed".

11. Describe fully how you would carry out a calibration test on a Bourdon pressure gauge.

Sketch a typical calibration curve and comment on possible errors and methods of correction.

12 Describe, with the aid of a clearly labelled diagram, the operation of a direct acting pneumatic relay.

State, within a measuring system, where such a device is usually used and give the reason for its use.

13. Explain, with the aid of sketches:
  (a) an electrical method,
  (b) a non-electrical method,
of measuring fluid flow, listing advantages and disadvantages of each method.

14. What is meant by the term "system response"? Give three examples of system response.
How is the "time constant" or "system lag" measured?
Why is it beneficial in a control system to reduce the time constant of that system?

15. Explain carefully, with the aid of a clearly labelled diagram, the principles of operation of either:
  i. a motion balance transmitter,
  or ii. a force balance transmitter.
Give two examples where either i. or ii. may be found in marine practice.

16. Using U-tube manometer diagrams, explain how:
  (a) absolute pressure,
  (b) gauge pressure,
  (c) differential pressure,
is measured.
Explain briefly how one such measurement may be transmitted to a remote recording station.

17. Explain, with the aid of sketches, the principle of operation of a nozzle-flapper system. Show graphically the relation between flapper clearance and output pressure. What reasons are there for fitting a feedback bellows in the system?

18. Sketch and describe a strain gauge.
What is the principle of operation and for what purpose can the strain gauge be utilised?

19. Describe what is meant by ramp, step and sinusoidal response when applied to a system. Illustrate each with a simple sketch.

20. Sketch a purge or "bubbler"system which could be used to measure the level of fluid in a tank.

What other information regarding the fluid would be necessary in order to infer a mass measurement from a level measurement?

How could such an instrument be developed to measure the draught of a ship?

21. (a) Draw a circuit diagram to show how junction diodes may be used to give full-wave rectification from a single-phase ac supply.

(b) A *npn* transistor is to be connected in common emitter mode for use in voltage amplification. Draw a simple circuit diagram to illustrate the configuration.

22. Give an example of level measurements which utilise the following physical principles:

(a) hydrostatic head,
(b) float movement,
(c) displacement.

Illustrate your answers with suitable diagrams and descriptive data.

23. Change of fluid level in a tank can be detected and measured by the change of capacitance of a capacitance probe. Sketch such a system and describe its operation in detail, from change in level to change in indication of measured level.

24. Make a clearly labelled circuit diagram of a "bridge" thermometer which has ambient temperature compensation, and also has provision for "zero or standardising" selection.

Explain the operating sequence followed to make the zero/standardising test before using the thermometer.

Describe how the ambient temperature compensation functions.

SPECIMEN EXAMINATION QUESTIONS

HNC (BTEC & SCOTVEC)

1. Describe, with the aid of a suitable diagram, the constructional details and operational working principle of one of the following devices:
  (a) oil mist detector,
  (b) torsionmeter,
  (c) oxygen analyser.

2. Draw a diagram of a system for fully automatic main engine jacket temperature control, incorporating steam preheating of the coolant, utilising split range control.
Comment on the type of control action to be used and the "fail safe" arrangement adopted.

3. (a) Construct a "truth table" for a two input NOR logic gate.
   (b) Sketch the circuit symbols for a *pnp* and *npn* transistor and clearly label the leads.
   (c) Draw a block diagram for part of a marine data logger and describe in detail the function of one particular block unit.

4. Automatic control valves are available with the following features:
  (a) linear or equal percentage characteristic,
  (b) power to open or power to close,
  (c) single or double seated.
Discuss the factors that determine the choice of each of the above features.

5. Make a diagrammatic sketch of a three term controller suitable for use in a feed water control system. The sketch should include measure and regulating units.

Briefly explain why it is necessary to incorporate:

      i. a negative feedback arrangement,

      ii. a relay (amplifier).

6. (a) State three advantages which solid state devices have compared to thermionic valves.

Why is silicon preferred to germanium for use in solid state rectifiers?

    (b) State the function of the gate in a transistor.

Sketch waveforms of ac supply voltage, load voltage and gate pulse for a thyristor when the trigger angle is 90°.

7. Make a diagrammatic sketch of a steam flow plant incorporating a flowmeter utilising an orifice plate and a differential pressure transmitter.

Explain the operating principle of the transmitter.

Sketch the output signal graph with and without square root extraction.

8. Explain the meaning of the following terms used in control terminology:

      (a) offset,

      (b) fail safe,

      (c) proportional band,

      (d) derivative action time,

      (e) cascade control.

9. Explain, with the aid of a diagram, a controller utilising proportional plus integral action.

Describe how such a controller is "tuned" within a system.

10. A boiler pressure is to be maintained at 5 bar by a pneumatic proportional controller. The pressure element has a range from 3 bar to 6 bar and the proportional band is set at 20%.

(a) Calculate maximum and minimum pressures corresponding to no load and to full load. (5·3 bar, 4·7 bar).

(b) If the controller output pressure range is from 1·2 bar to 2·0 bar, draw a graph of boiler pressure and controller output pressure, and from the graph determine the boiler pressure at a controller output pressure of 1·44 bar. (5·16 bar).

11. (a) Sketch typical forward and reverse characteristics for a zener diode and briefly explain the action of this device when a reverse voltage is applied to it.

(b) State one application of the zener diode in marine equipment.

(c) A 10V, 500 mW zener diode used as a voltage stabiliser is supplied at 40 V through a series resistor of 500 ohms. If the load voltage is stable for diode currents greater than 5 mA, draw the circuit diagram and calculate:

i. the maximum and minimum load current for stable operations, (55 mA, 10 mA),

ii. the minimum power rating of the series resistor. (1800 mW).

12. Describe, with the aid of sketches, the construction and operational features of one of the following devices:

(a) a buoyancy tube liquid level transmitter,

(b) a pneumatic differential pressure transmitter,

(c) a vapour pressure thermometer utilised for remote indication.

13. With the aid of simple diagrams illustrate the response of a proportional controller, when the system is subject to a load change if:

i. the proportional band is too narrow,

ii. the proportional band is too wide,

iii. the proportional band is at the optimum setting.

Briefly describe why a relay valve is used with such a controller and state what secondary function relay valves usually perform.

14. An electric torsionmeter, shown in Fig. G is fitted to the output shaft of a marine engine.

(a) With the aid of this diagram explain in detail the principle of operation.

(b) Explain how the instrument would be used to determine the output power using the appropriate formula. Detail carefully how shaft constants are obtained.

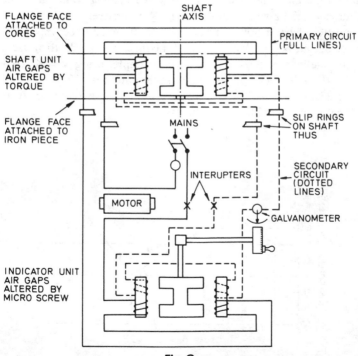

Fig. G

15. Sketch a thyristor speed control system for a dc motor. Discuss briefly how the system will respond to an increase in the speed set point potentiometer and explain the motor protection features included in the system.

16. (a) A reverse-acting two term pneumatic controller is subject to an input which varies as shown in Fig H.
Construct a diagram showing the variation in controller output due to:
      i. proportional action,
      ii. integral action,
      iii. combined (total output).

   (b) State whether the diagram constructed indicates that the load on the plant has been restored to the original value after the disturbance and give a brief reason for your answer.
Proportional band 80%, integral action time 120 seconds, controller output before disturbance 70 kN/m². (K_1 = 1·25 then use a tabular method; at three minutes, for example, using $V = K_1 (\theta + K_2/K_1 \int \theta \, dt)$ the total change in controller output is −37·5, i.e −37·5 = − 1·25 (20 + $\frac{1}{2}$ × 20). Controller output is not restored to its original value so neither is the load).

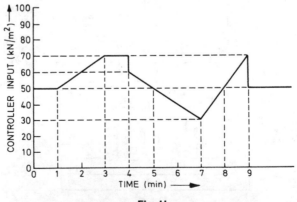

Fig. H

17. Fig I shows the arrangement of a spray-type of attemperator in which the outlet steam temperature is controlled by varying the amount of feed water injected (1 kg feed is required to attemperate 10 kg of steam).

Assuming that the system has been adjusted so that the measured variable (outlet steam temperature) is at the desired value when the load is 40 kg of steam per minute determine:

    i. the offset if the load changes to 20 kg of steam per minute, (−16°C),

    ii. the limits of load that can be accommodated, assuming that the offset shall not be more than ± 30°C. (2·5–77·5 kg/minute).

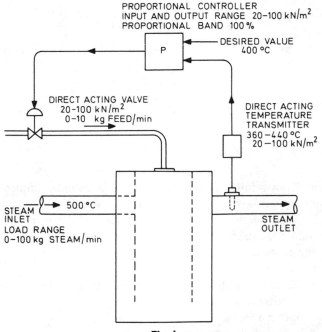

Fig. I

18. (a) Sketch, and state, the type of logic gate which obeys the following:
"If all inputs are one the output is zero; for all other combinations the output is one." (NAND).

(b) Fig. J (left hand sketch) shows the circuit of an electrical logic gate. By inspection or otherwise, state what type of gate the circuit represents and write down the equation for the output at Z. (OR, A + B + C).

(c) Construct a truth table for the logic circuit shown in Fig. J (right hand sketch) and give the outputs at Z from each logic unit. (For example, ABC conditions 011, outputs 01100).

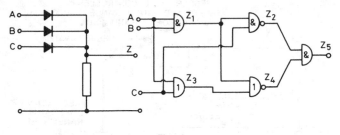

Fig. J

19. (a) Draw a fully labelled diagram of a pneumatic temperature transmitter operated by a liquid or gas filled sensor.

(b) Describe the operation of the transmitter in response to a temperature increase, making particular reference to the need for, and operation of, negative feedback.

20. (a) Define, and illustrate with simple diagrams, the following:
    i. distance-velocity lag,
    ii. transfer lag,
    iii. potential correction.

(b) Define, and state the mathematical representation of the following modes of control:
    i. proportional,
    ii. integral.

21 (a) Draw a system for the automatic control of main turbine engine gland steam pressure. The system is to use two control valves, one make up and one rejection operating on an equal split range with 4 kN/m² underlap in the signal.

(b) State, or show on the system drawing:
    i. the control valve actions and fail safes,
    ii. the operating signal range for each valve,
    iii. the reason for underlap in the signal.

(c) Describe clearly the actions throughout the control loop as the gland steam pressure rises.

22. An oily water separator is fitted with two teflon coated capacitance probes, each with separate measuring bridge and output:
    i. an oil depth probe vertically, to operate the oil discharge valve,
    ii. an interface probe horizontally, to operate an alarm and oily bilge pump trip.

(a) Sketch the arrangement and explain clearly in terms of capacitance change, the operation of each probe as oil level varies.

(b) If the separator is completely empty all alarm conditions will operate when starting up. Explain the reason for this.

23. (a) In a calibration check on a $P + D$ controller the ramp input is increased linearly at 1% per minute. This produces an immediate 4% step change in output, after which the output changes linearly at 2% per minute.

Sketch input and output characteristics and determine:
    i. derivative action time, (2 minutes),
    ii. proportional band. (50%).

(b) A step change of 6% is applied to the input of a $P + I$ controller and the output undergoes a sudden step change of 4% and after a time interval of 2 minutes the total output change is 10%.

Sketch input and output characteristics for this calibration test and determine:
    i. proportional band, (150%),
    ii. integral action time. (1·333 minutes).

(c) Define the terms distance-velocity lag and exponential lag.

24. Fig. K shows a boiler arranged for manual control of the drum steam pressure.

(a) Complete the diagram by adding the necessary components and connections to convert to an automatic closed loop combustion control system.

(b) Describe the operation of the automatic system.

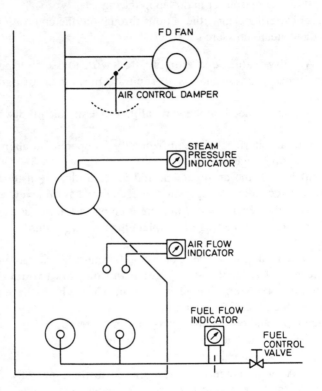

Fig. K

SPECIMEN EXAMINATION QUESTIONS

HND (BTEC & SCOTVEC)

1. With the aid of a diagrammatic sketch describe a type of controller which incorporates proportional, integral and derivative actions. Explain clearly how each action is generated and comment briefly on possible interaction that may occur.

2. Sketch circuit diagrams of resistance-capacitance networks which will provide an output signal that approximates to:
   (a) proportional plus derivative of input signal,
   (b) proportional plus integral of input signal.
Show, in each case, how these networks could be incorporated into a closed loop control system so as to improve the operating performance of the system.

3. For a closed loop position control system:
   (a) make a sketch to illustrate how the components are connected,
   (b) describe the sequence of events following movement of the input shaft,
   (c) explain the term "overshoot" in relation to such a system,
   (d) describe how damping feedback could be incorporated.

4. (a) Define the terms proportional and integral action.
   (b) Sketch the open loop characteristic response for proportional and integral action when a step input change is applied in each case.
   (c) Calculate the integral action time if a step input change of 0·04 bar applied to a pneumatic controller set with a 100% proportional bandwidth gives a response change of 0·2 bar in 3s. (0·75 s).

5. (a) Explain, by means of a block diagram or otherwise, the fundamental principle of a dc chopper amplifier.

(b) Sketch a circuit/block diagram to show how a chopper amplifier is used in an instrument servo-mechanism which records dc potentials by means of a self balancing potentiometer. Briefly describe the operating principles utilised.

6. (a) Explain by means of a two transistor analogy, or otherwise, how a gate input pulse will cause a transistor to conduct.

(b) A thyristor stack is used to regulate the heating power to a cargo hold in order to provide automatic temperature control. Draw a block diagram to show the basic components for such a control system.

(c) Using switches and a lamp as an example, explain with the aid of simple diagrams what is meant by the following logic terms:
   i. the AND function,
   ii. the OR function.

7. With the aid of a diagram describe the construction and operating principle of a valve positioner.
Give three reasons why such a device may be utilised in a control system.
Explain how a positioner may be adjusted to alter the valve stroke from 1·2-1·8 bar to 1·3-2·0 bar and state why gain should be as high as possible.

8. Explain the principles of viscosity measurement and detail the type of instrumentation used.
Sketch an oil viscosity control system and describe the operating principle.
Give reasons for which control actions you would incorporate in the controller.

9. Describe with a block diagram the operational construction of a data logger suitable for marine use.
Clearly indicate the functions of alarm annunciation recording of input signal information, and analogue to digital conversion.

10. (a) Describe the effect of:
     i. positive feedback,
     ii. negative feedback,
on a closed loop speed control system

    (b) For voltage at input e_1, output e_0 and error actuating input e to a negative feedback electronic amplifier of forward gain G and negative feedback fraction F, derive the following:

$$\frac{e_0}{e_1} = \frac{G}{1 + FG}$$

$$\frac{e}{e_1} = \frac{1}{1 + FG}$$

If the "open loop" gain is infinite, determine F if the overall gain is 25. (0·04).

11. (a) Draw a block diagram of a remote position control servo-mechanism suitable for controlling the angular position of the ship's rudder. Clearly label the inputs and outputs for each block and use them to explain what is meant by proportional control.

    (b) Explain with the aid of diagrams why damping needs to be introduced into the system, when a step change of input is applied, and state one method of introducing damping into such a system.

12. (a) Explain the meaning of the following terms, using suitable diagrams where appropriate:
     i. distance-velocity lag,
     ii. transfer lag,
     iii. time constant,
     iv. thermal capacity.

    (b) A thermometer bulb is housed in a pocket. Show by means of a response curve an estimation of the effect of the pocket and state what factors influence the design of the pocket to minimise this effect.

13. (a) Briefly describe why it is desirable to employ a multi-element system in the control of boiler water level.

(b) Make a diagrammatic sketch of a three-element boiler water level control system, naming all the parts and giving a brief description of its operation.

(c) Sketch a diagram showing the variations in boiler water level and feed pump load that might be expected if a water tube boiler is subject to a step increase in load, with:

i. single-element control,

ii. three-element control.

14. (a) Fig. L shows the block diagram of a low power remote position control servo system with velocity feedback damping. State the function of each component in the system.

(b) Explain briefly how the system will respond to a step input and sketch the corresponding response curve for:

i. switch S open,

ii. switch S closed.

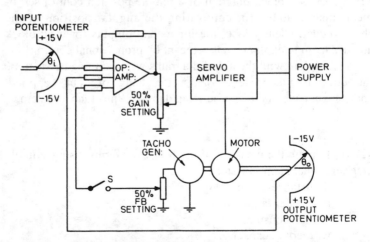

Fig. L

15. The air pressure control system shown in Fig. M has been adjusted such that at a load of 6 kg/min the equilibrium value of the controlled condition is at the desired value of 30 bar.
Determine:-

   i. the proportional control factor (μ) of the system if the proportional band of the controller is 50%, (0·5),

   ii. the offset if the load changes to 8 kg/min, (−4 bar),

   iii. the proportional band setting required such that for a load change from 6 kg/min to 2 kg/min the offset is limited to 2 bar, (12·5%),

   iv. the smallest load that the system can deal with if the proportional band of the controller is set to 100%, (3·5 kg/min),

   v. the proportional band of the controller such that the system can deal with any load within the range 0-10 kg/min. (41·7%).

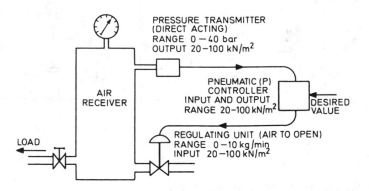

Fig. M.

16. (a) Sketch the forward characteristic for a thyristor and indicate clearly:
    i. the hold current level,
    ii. the effect of increasing the gate trigger pulse upon the forward break-over voltage.

(b) State three advantages of a thyristor compared to other types of controlled rectifier.

(c) Fig. N shows the circuit of an ac controller:
    i. explain briefly how the power in the load is controlled,
    ii. sketch the wave forms of load voltage and gate pulse for half maximum load power.

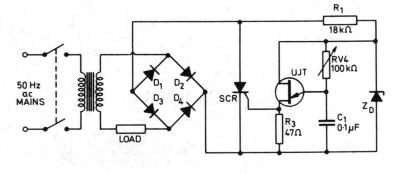

Fig. N

17. (a) Define the following terms:
    i. integral action,
    ii. derivative action.

(b) With the aid of suitable diagrams derive expressions for:
    i. integral action in terms of controller gain and integral action time,
    ii. derivative action in terms of controller gain and derivative action time.

18. Fig. O shows the arrangement of a damper position control system for a boiler employing damper control of the final steam temperature, and the variation in steam temperature, during a transient condition. Draw an accurate diagram showing the damper position during the same transient period. The damper may be assumed to be in the 60% open position immediately before the transient condition commences. (System proportional control factor μ = multiple of proportionality characteristics, coefficients, i.e. $2 = 2 \times 0.8 \times 1.25$ then use a tabular method; at two minutes, for example, using $\Phi = -\mu(\theta + I/S \int \theta \, dt)$ the damper is 20% open, i.e. $-40 = -2 (10 + 1/0.5 \times 5)$. Damper is closed between 2.5 and 5 minutes).

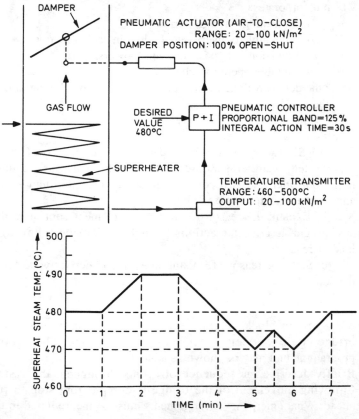

Fig. O

19. (a) Draw a detailed diagram of an instrument suitable for converting electrical control signals into pneumatic signals.

(b) Explain clearly the operation throughout the instrument in response to an increased input signal.

(c) Discuss briefly the relative advantages of electro-pneumatic control.

20. (a) Draw the circuit for a two input diode-transistor logic OR gate suitable for operating an alarm system. Construct the truth table and describe the operation of the gate.

(b) Construct a logic flow diagram for a sootblowing system which incorporates:

     i. a warming through period,
    ii. retractable blowers which blow in each direction,
    iii. provision to bypass selected blowers,
    iv. suitable operation checks.

Show the details for one blower only and draw a block diagram for the others.

21. (a) Sketch a fully labelled diagram of a two element superheated steam temperature control system which uses a three term controller and two control valves, a steam to attemperator and an attemperator bypass valve.

(b) Explain the reason for using two element control in this system and describe the actions throughout the loop following a load increase.

(c) State the reasons for using integral and derivative terms in the controller.

22. Show in block diagram form the basic requirement for a bridge control system for a main diesel engine. The control programme unit may be shown as a block.

Briefly describe the sequence of events throughout the system following movement of the bridge telegraph from stop to full ahead. What emergency arrangements must be incorporated in the system and fitted on the bridge?

23. Fig. P shows the main components of a jacket cooling water system for a marine diesel engine, outlet temperature is controlled condition.

(a) Assuming that it has been decided to modify the system to provide a split level, two element, cascade closed loop pneumatic control system, complete the diagram naming clearly all the additional components required.

(b) Describe the operation of the modified system giving typical operating ranges and stating clearly the modes of action employed *e.g.* direct acting proportional.

(c) Describe the effects on the system of:
    i. an increase in sea water temperature,
    ii. a decrease in engine load.

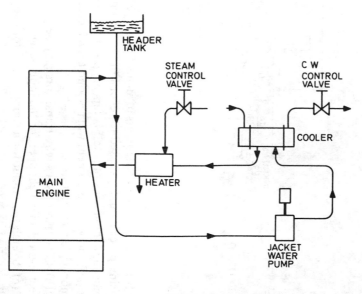

Fig. P

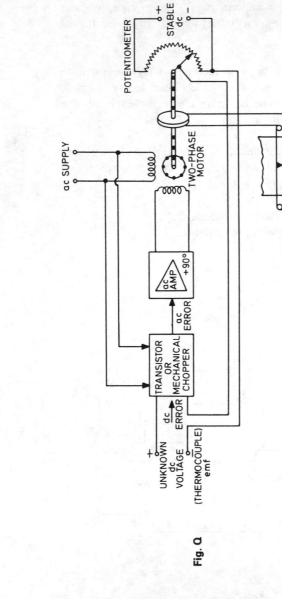

Fig. Q

24.   (a) Fig. Q shows the elements of an instrument servo – self balancing potentiometer – used to record the temperature of a thermocouple by means of a self balancing potentiometer. Discuss briefly how the instrument will respond to an increase in the thermocouple temperature.

(b) Give two reasons why transistors may be preferred to an electro-mechanical chopper for the input stage.

(c) Sketch the circuit for a simple transistor chopper and briefly describe its operation.

INDEX

REED'S MARINE ENGINEERING SERIES

The series covers the full range of subjects for all grades of the Department of Transport Certificates of Competency in Marine Engineering (administered by SCOTVEC). The books will be extremely useful for marine engineer cadets and other engineering students studying on Business and Technician Education Council (and SCOTVEC) Courses. Material is presented from first principles with many diagrammatic sketches and worked solutions to examples.

Vol. 1. MATHEMATICS
Vol. 2. APPLIED MECHANICS
Vol. 3. APPLIED HEAT
Vol. 4. NAVAL ARCHITECTURE
Vol. 5. SHIP CONSTRUCTION
Vol. 6. BASIC ELECTROTECHNOLOGY
Vol. 7. ADVANCED ELECTROTECHNOLOGY
Vol. 8. GENERAL ENGINEERING KNOWLEDGE
Vol. 9. STEAM ENGINEERING KNOWLEDGE
Vol. 10. INSTRUMENTATION & CONTROL SYSTEMS
Vol. 11. ENGINEERING DRAWING
Vol. 12. MOTOR ENGINEERING KNOWLEDGE

These books are obtainable from all Nautical Opticians, Chartsellers and Booksellers, or direct from:

THOMAS REED PUBLICATIONS LIMITED
Weir House, Hurst Road, East Molesey,
Surrey KT8 9AQ, U.K.